中国戏曲学院『十三五』教材系列

中英文对照京剧行当术语

编著 王立新 阙艳华 周欣 苏凤 田芸

校译 [英] 陶西雷（Ashley Thorpe）

文化艺术出版社
Culture and Art Publishing House

U1003202

图书在版编目（CIP）数据

中英文对照 京剧行当术语 / 王立新等编著. -- 北京：
文化艺术出版社, 2019.10
ISBN 978-7-5039-6797-9

Ⅰ. ①京… Ⅱ. ①王… Ⅲ. ①京剧—术语—汉、英
Ⅳ. ①J821-61

中国版本图书馆CIP数据核字(2019)第213270号

中英文对照　京剧行当术语

编　　著　王立新　阚艳华　周　欣　苏　凤　田　芸
校　　译　[英]陶西雷（Ashley Thorpe）
责任编辑　张　恬　张月峰
书籍设计　丁智睿　赵　矗
出版发行　文化艺术出版社
地　　址　北京市东城区东四八条52号（100700）
网　　址　www.caaph.com
电子邮箱　s@caaph.com
电　　话　（010）84057666（总编室）　84057667（办公室）
　　　　　（010）84057696—84057699（发行部）
传　　真　（010）84057660（总编室）　84057670（办公室）
　　　　　（010）84057690（发行部）
经　　销　新华书店
印　　刷　鑫艺佳利（天津）印刷有限公司
版　　次　2019年10月第1版
印　　次　2019年10月第1次印刷
开　　本　710毫米×1000毫米　1/16
印　　张　8.5
字　　数　120千字
书　　号　978-7-5039-6797-9
定　　价　68.00 元

版权所有，侵权必究。如有印装错误，随时调换。

中国戏曲学院
National Academy of Chinese Theatre Arts

中国戏曲学院"十三五"教材系列
编委会

主　任　冉常建

委　员　（按姓氏笔画排序）

于建刚　马　路　王志苹　王绍军　乔慧斌　李　钢　李　威
张　尧　张文振　梁建明　舒　桐　谢振强　谭铁志　颜全毅

秘　书　刘　芳

目录

第二章　旦行 (Dan)

第三章 净行 [Jing (painted face)]

第四章 丑行 (Chou)

行当概述

行当，戏曲专业术语。是中国戏曲特有的一种表演体制。简称“行”。

据曾白融主编的《京剧剧目辞典》，京剧的剧目共计5300出，保留的剧目千出以上。这上千出剧目都是由不同的角色共同演绎、创造的。角色按照性别、年龄、职业、身份、地位、形象特征等诸多因素分别由不同的演员完成，形成了京剧的行当。

戏曲的功效源于戏剧，从祭祀、祈神到自娱、娱人、育人，其中蕴含着情感的宣泄和文化的传递。为了实现其功能，仿效生活是必由之路。生活是个体差异巨大的人的各不相同的喜怒哀乐，在真善美、假恶丑交织的多样化世界的重复和演进。戏曲艺术通过语言、音乐，借助叙事框架，展现人物、事件。描摹生活，展现情趣，针砭时弊，惩恶扬善，援古证今，鉴往知来，戏曲以此不断完善、提升着自身的艺术性和社会功效。

中国戏曲的发展在经历了原始戏剧和初级戏剧阶段后进入了成熟期，即宋朝的杂剧时代。宋杂剧成熟的主要特征是角色行当的出现，包括末泥、引戏、副净、副末、装孤五个角色。到了清朝，戏曲的角色行当日臻丰富、完善。按照乾隆时期戏曲作家李斗撰写的《扬州画舫录》记载：“梨园以副末开场，为领班。副末以下，老生、正生、老外、大面、二面、三面七人，谓之男脚色；老旦、正旦、小旦、贴旦四人，谓之女脚色；打诨一人，谓之杂。此江湖十二脚色。”这种表演分工的格局，形成了中国最古老的剧种之一——昆曲的“脚色行当”的基本体制，并对京剧行当的形成产生了很大影响。19世纪中叶到20世纪初，包含徽剧、汉调、昆曲、梆子、京腔等地方戏曲元素，集中国戏曲之大成的新的声腔剧种——京剧登上了京城舞台。在长期艺术实践过程中，在借鉴、吸收其他古老剧种成就的基础上，京剧创造出自身的表演规范、艺术语汇、程式体系，尤其在人物塑造方面发展、完善了生、旦、净、丑四大人物表演行当，使其各具独特的艺术吸引力。

戏曲划分行当以后，对于演员个体来说，可以按照行当属性、特点、要求等，在一定范围内集中学习，通过对演唱、表演等方法、技巧的掌握及运用，与其他角色一起，借助舞台，完成戏曲的娱乐、教化等功能，从而最终实现京剧自身的文化、艺术价值。

通过具体行当的学习，演员可在掌握唱、念、做、打等基本功的前提下，根据自身嗓音、外形等内在、固有的特点，进行艺术创作。借助化妆、服装等方面的艺术处理，演员还可以发挥自身创作潜力，打破年龄、性别的藩篱，展现戏曲艺术独特、无限的想象力和创造力。在观看戏曲节目时，观众既可以欣赏到演员的本色表演，又可以逆势反观。例如，生活中秀美的孟小冬可以是舞台上美髯飘动的义士程婴，抑或是满腹经纶的诸葛亮；铁面包公浑然刚劲，窦尔敦气度雄武，而他们的扮演者是女艺术家齐啸云；现实中洒脱正义的梅兰芳在舞台上幻化为袅娜迷人的贵妃；蔡文姬、窦娥等鲜活的舞台女性形象被只身痛打伪警察的程砚秋演绎得淋漓尽致、惟妙惟肖；稚嫩的娃娃可以装扮成古稀的佘太君，年逾七旬的叶少兰举手投足就是少壮英武的吕布。艺术家游刃于巨大的创作时空，所创造的戏曲人物如宇宙繁星，熠熠生辉。京剧艺术的魅力通过行当的划分得以充分彰显。结合“唱、做、念、打”，人物、剧情通过舞台表演得以具象化。行当之间、演员之间的相互沟通、借鉴又为京剧艺术的提炼、升华提供了可能。

戏曲行当的划分对于观众来说，仿佛是一把打开艺术宝库的钥匙，在辨别角色、了解剧情、咀嚼唱词、品味唱腔的基础上，陶醉于名家大师的艺术风采，尽享京剧艺术的美妙与深邃，进而激发了对博大精深的中国传统文化由衷的挚爱。

人作为天地之精华，万物之灵长，上下求索，生生不息：阡陌百姓为生计奔波忙碌；浪漫恋人为情爱欢愉烦恼；将帅士卒为忠义鏖战厮杀；智士贤达为人生思考探究。他们的酸甜苦辣、悲欢离合、真假虚实被一方小小的京剧舞台呈现着、渲染着、传承着……依托行当进行表演的京剧以其独有的艺术助推力向世人传达着中华民族独特的文化内涵、精神追求、审美取向和道德理想。

A Survey of Role Types

A "role type" (known as "hang dang", or just "hang", in Chinese theatre circles), is a unique system in performance.

According to the Dictionary of the Repertoire of Peking Opera, edited by Zeng Bairong, altogether there have been 5300 plays in Peking Opera's history, of which more than 1000 are still performed today. All of these are performed by various role types, distinctly grouped by gender, age, occupation, social status, and character trait.

Chinese opera's efficacy stems from its wider relationship to other dramatic forms: from ceremonies of worship, the offering up of performances to entertain the gods, performances staged for cathartic entertainment, and the education of the populace via the transmission of culture. To achieve these aims, it was necessary to emulate people and their daily lives. Life consists of a rich tapestry of individuals: there are the good who feel anger, sorrow, and joy, and they live amongst the liars, the wicked and the shameful. Through the use of language and music, drama narrated the changes of the populace and their environment. Here, life was described, taste was depicted, the disadvantaged were highlighted, evil was punished and good was rewarded. With the help of the past, the present was confirmed, and the future was predicted, and through this opera enhanced its own artistic and social function.

After these primitive periods in dramatic history, drama matured in the Song dynasty, due to the emergence of Song Zaju, a kind of comic skit. This maturity coincided with the appearance of five basic role types, including MONI, YINXI, FUJING, FUMO and ZHUANGGU. These role types were developed and perfected in the Qing Dynasty. The Record of Pleasure in Yangzhou, written by Li Dou, a dramatist in the Qianlong

Reign, noted how a "piece began with the FUMO who opened the play. Together with the FUMO, there were six other male roles, including LAO SHENG, ZHENG SHENG, LAO WAI, DA MIAN, ER MIAN, SAN MIAN. The piece also featured four female roles, namely LAO DAN, ZHENG DAN, XIAO DAN and TIE DAN. In addition, there was a clown, called ZA. These roles are generally known as the Twelve Roles of Opera." These role types formed the basis for Kun Opera, one of the oldest Chinese Opera genres, which in turn, influenced the formation of role types in Peking Opera. From the mid-nineteenth to the beginning of the twentieth century, nourished by local theatre forms such as Hui Opera, Handiao Opera, Kun Opera, Bangzi Opera and Jingdiao Opera, Peking Opera, a new kind of music theatre, made its entrance on Beijing's stage, the capital of the Qing Dynasty. During the long course of its artistic practice, Peking Opera never stopped learning from and absorbing the achievements of other Chinese operas. Peking Opera established new acting forms and styles, artistic terms, and systems of programming. This was especially the case with characterisation, where four basic role types, Sheng, Dan, Jing and Chou, were distilled so that each had their own unique artistry.

After the division of acting roles, individual actors could concentrate on learning the specific requirements, attributes and characteristics of their role type. Through the mastery of the singing and performance skills required of their role, and by combining actors of different role types together, a complete entertainment was created to amuse and educate audiences, enabling Peking Opera to achieve its unique cultural and artistic status.

Through the study of specific role types, performers can understand how to deploy the basic skills of singing (chang), recitation (nian), dance-acting (zuo) and fighting (da), and shape them according to their own physical characteristics, such as their own vocal qualities and physical appearance. Through makeup and costumes, they can also produce brand new stage characters that transcend their own age and gender, demonstrating the infinite imaginative creativity of this traditional art.

While enjoying performances, audiences may see the real actor or actress behind the character. Meng Xiaodong, a delicate lady in real life, appeared on stage as Cheng Ying, a righteous, bearded, male military adviser. The characters of Bao Gong, a well-known upright official in Song Dynasty, and Dou Erdun, a heroic outlaw in the Qing Dynasty, were both performed by another female performer, Qi Xiaoyun. The actor Mei Lanfang transformed himself into female characters such as imperial concubines. Talented female scholars and other traditional female characters were also successfully impersonated in different plays by the male actor Cheng Yanqiu. A young child might play an old-aged matriarch, or the famous actor Ye Shaolan may play the part of a valiant young general on stage even though he is over seventy years old. Artists are more than equal to the task of creating a glittering array of colourful characters.

The charm of Peking Opera is aptly demonstrated by the division of role types. Combined with the basic skills of singing, recitation, dance-acting and fighting, characters and storylines are vividly realized in performance. Actors learn from each other across the different role types, sharing ideas and techniques to further refine the art.

For audiences, the division of role types is the key to opening the treasure house of Chinese theatre arts. They can easily distinguish between the various roles, comprehend the meaning of the story, mull over the meaning of lyrics, and enjoy the music from the arias. Everyone becomes enchanted by the skilful dexterity of these artistic masters, and enjoys the beauty and the profound meaning of Peking Opera. This inspires us to love our great Chinese culture from the bottom of our hearts.

The evolution of society is the result of human development. As the essence of the universe, the soul of all things, humanity never ceases to explore the unknown. People struggle to support their families; true lovers are embroiled in the pitfalls and pleasures of romance; warriors make sacrifices as they fight for victory; the wise consider their implacable fate. All of these things are depicted, glorified and magnified on the stage. With its unique theatrical appeal, Peking Opera, a performance based on the division of role types, has artistically conveyed the cultural essence, spiritual pursuit, aesthetic concerns and moral ideals of the Chinese nationality to the people of the world.

第一章　生行

生行

生行，戏曲表演的主要行当之一，除净、丑以外的男性角色都属于生行。生扮演男性人物。按照所扮人物的年龄、身份、性格特征和表演特点，大致分为老生、小生、武生等门类。生的面部装扮除了红生及勾脸的武生之外，大都采用素脸，也就是行内说的“俊扮”。

Sheng

Sheng is one of main types of roles in Peking Opera. It is one of the three main types of male role, the others being Jing and Chou. According to the age, identity and personality of the characters, Sheng roles can be further divided into categories such as Lao Sheng (aged male roles), Xiao Sheng (young male roles) and Wu Sheng (martial male roles). For the most part, the facial makeup of the Sheng role is simple and light, known to professionals as "Junban" (handsome) make-up. Exceptions include Hong Sheng (a type of Lao Sheng – older male role – with red facial makeup) or certain roles of the Wu Sheng (martial male) role.

老生

1.

老生又名正生，因戴髯口，又称须生，俗称胡子生，是京剧生行的一种。扮演中年以上成熟稳重、性格正直刚毅的正面男性人物。化妆都要戴胡子（髯口），其颜色分黑、白、灰三色。三绺黑胡子，术语称“黑三”，灰色的叫“黪（cǎn）三”，白色的叫“白三”。整片满口的胡子，不分绺，术语称“满”，有黑满、黪满、白满。老生造型庄重端方、器宇轩昂。老生一般重唱功，唱和念白都用本嗓（真嗓），声腔以西皮、二黄为主，板式丰富多样。唱腔平稳舒展、苍劲流畅，念韵白。做功方面沉稳从容，多运用身段、步法，结合眼神和面部表情，表现剧情及人物特定的处境。

老生一般分为文老生、武老生两类。按照表演的侧重可分为唱工老生、做工老生、靠把老生、箭衣老生；如果按照人物的身份、地位和所穿戴的服装来划分，可以分为王帽老生、袍带老生、褶子老生、靠把老生和箭氅老生等。

王帽老生：头戴王帽，身穿王袍的帝王角色，如《打金砖》中的刘秀。

袍带老生：头戴纱帽，身穿蟒袍或官衣的官员角色，如《群英会》中的鲁肃。

褶子老生：穿褶子服（便服）的绅士及在野官员，如《捉放曹》中的陈宫。

这里的“褶子”指传统戏曲服装，在《辞海》里的释义为：“剧中古代平民的便服和帝王官绅的衬衣。一般用绸制成。男式褶子大领大襟，有水袖，分花色、素色两种。花色褶子以彩线绣鸟兽花草等图案；素色褶子以黑色、蓝色者为多。女式褶子长仅及膝。青年、中年妇女所着者小领小襟，老年妇女所着者大领大襟。”由于戏曲表演者口口相传，习惯上把“褶子”读为“xué”，音同“学”，与《辞海》上注音“xízi”不同。

老生的流派有孙（菊仙）派、谭（鑫培）派、余（叔岩）派、言（菊朋）派、高（庆奎）派、麒（麟童）派、马（连良）派、谭（富英）派、杨（宝森）派、奚（啸伯）派等。

Lao Sheng

The Lao Sheng (aged male roles), is also known as Zheng Sheng (principal male role), as well as Xu Sheng or more commonly Huzi Sheng (male roles with a beard). The Lao Sheng usually plays middle-aged male roles or respected and steady characters of integrity in Peking Opera. The Lao Sheng role normally wears a rankou (artificial beard) in performance. The colours of rankou are black, white, and grey. The term heisan (literally "black three") is used to denote a form of black beard that consists of three separate strands, the cansan ("grey three") for a grey beard with three strands, and baisan ("white three") for a white beard with three strands. The term man (literally "full") is used for a full beard that does not have strands: heiman for full black beards, canman for grey ones, baiman for white ones. The deportment of the Lao Sheng on the stage is serious and dignified. The technique of singing in Lao Sheng roles is particularly important. The natural voice (true voice) is used both in singing and in the spoken parts when performing. The tunes for the Lao Sheng are sung mainly in the Xipi and Erhuang modes (the two chief types of musical mode in Peking Opera, roughly equivalent to major and minor in Western music) and in different types of beat form and rhythmic structure. The tunes of Lao Sheng are steady and smooth, and the spoken parts are rhythmical. The movements of Lao Sheng are sedate and calm, exhibiting the particular narrative arc of the character through body gestures and gaits, combined with expressive eyes and countenance.

The Lao Sheng is generally divided into two types: Wen Lao Sheng (civilian aged male roles) and Wu Lao Sheng (military aged male roles). It also can be classified into Changgong Lao Sheng (aged male roles that emphasize singing), Zuogong Lao Sheng (aged male roles that emphasize acting) according to the emphasis of the role. In addition, the Lao Sheng can be further differentiated according to the clothing they wear to indicate their status, such as Wangmao Lao Sheng (aged male role who wears a "Wangmao", a crown), Paodai Lao Sheng (aged male role who wears a "Mangpao", a court robe), Xuezi Lao Sheng (aged male role who wears a "Xuezi", a casual coat), Kaoba Lao Sheng (aged male role who wears a "Kao", an armour), and Jianchang Lao Sheng (aged male role who wears a "Jianchang", an archer's cloak).

The Wangmao Lao Sheng (aged male role who wears a crown) plays imperial roles, and wears a Wangmao (emperor's crown) and a Wangpao (emperor's robe), such as Liu Xiu (emperor of the Eastern Han Dynasty) in "Beating the Emperor with a Gold Brick" (Da Jinzhuan).

The Paodai Lao Sheng (aged male role in a court robe) portrays civil officials and wears a Shamao (gauze cap) and a Mangpao (court robe) or Guanyi (official clothes), such as Lu Su from "A Meeting of the Heroes" (Qun Ying Hui).

The Xuezi Lao Sheng (aged male role in a casual coat) plays gentlemen or officials who are out of office, and wears a Xuezi (casual coat), such as Chen Gong from "The Capture and Release of Cao Cao" (Zhuo Fang Cao).

"Xuezi" here refers to costumes in traditional Chinese theatre. It is customary for the theatre masters to teach pupils orally which inspires true understanding within. That is to say, theatrical customs are passed from one generation to the next. With the passage of time, the dress is called "Xuezi", which is different with Ci Hai (a scale Chinese dictionary). In the dictionary, "xizi" is the correct pronunciation.

The different artistic schools (or styles) of Lao Sheng include Sun (Juxian), Tan (Xinpei), Yu (Shuyan), Yan (Jupeng), Gao (Qingkui), Qi (Lintong), Ma (Lianliang), Tan (Fuying), Yang (Baosen), and Xi (Xiaobo).

1.1.

唱工老生

唱工老生也叫作安工老生，以唱为主，动作幅度较小，神态闲适从容，沉着安稳，如《捉放曹》中的陈宫、《文昭关》中的伍子胥、《辕门斩子》中的杨延昭、《空城计》中的诸葛亮、《四郎探母》中的杨延辉、《李陵碑》中的杨继业等。

Changgong Lao Sheng

Changgong Lao Sheng (aged male roles that emphasize singing) is also known as Angong Lao Sheng (sedate aged male roles), and the actor concentrates on singing in the performance, with a sedate and calm deportment and slight movements. Examples of this role include Chen Gong in "Capture and Release of Cao Cao" (Zhuo Fang Cao), Wu Zixu in "Wenzhao Pass" (Wenzhao Guan), Yang Yanzhao in "Executing A Son At Yuanmen" (Yuanmen Zhan Zi), Zhuge Liang in "Empty-Town Ruse" (Kong Cheng Ji), Yang Yanhui in "Si Lang Visits His Mother" (Si Lang Tan Mu), and Yang Jiye in "The Tombstone of Li Ling" (Li Ling Bei).

唱工老生——《空城计》中的诸葛亮

做工老生

1.2.

做工老生又称衰派老生、做派老生，是以表演为主的一种行当。衰派老生也有两种解释：一种解释，这类老生专门扮演年老体弱、戴白胡子的角色，所以叫作衰派。例如《清风亭》中的张元秀、《扫松下书》中的张广才、《三娘教子》中的薛保、《马义救主》中的马义、《徐策跑城》中的徐策等。另一种解释，因为以做功为主，一般都是表演精神上受了刺激，情绪紧张、神态粗犷、动作激烈的角色，这些角色的精神状态近乎颓唐衰朽，所以叫作衰派。例如《坐楼杀惜》中的宋江、《打棍出箱》中的范仲禹、《失印救火》中的白槐、《春秋笔》中的驿丞张恩、《战蒲关》中的刘忠等。

Zuogong Lao Sheng

Zuogong Lao Sheng (aged male roles that emphasize acting), is also known as Shuaipai Lao Sheng (aged male roles in decline) and Zuopai Lao Sheng (expressive aged male roles), and focuses mainly on acting. It bears these names for two reasons. The first is that this type of role often portrays elderly characters who wear a white beard and are generally in poor health, such as Zhang Yuanxiu in "Qingfeng Pavilion" (Qingfeng Ting), Zhang Guangcai in "Delivering A Letter" (Sao Song Xia Shu), Xue Bao in "Sanniang Gives Her Son a Lecture" (Sanniang Jiao Zi), Ma Yi in "Ma Yi Rescues His Master" (Ma Yi Jiu Zhu), and Xu Ce in "The Great News" (Xu Ce Pao Cheng). The second reason is that because this type of Lao Sheng emphasizes acting as the main element of performance, the characters portrayed experience high emotions, nervousness, have an unrefined demeanour, and are agitated. Examples include Song Jiang in "Song Jiang Kills Yan Xijiao" (Zuo Lou Sha Xi), Fan Zhongyu in "Sticks Fall Out the Trunk" (Da Gun Chu Xiang), Bai Huai in "Loss and Recovery of the Official Seal" (Shi Yin Jiu Huo), Zhang En in "The

Honest Pen" (Chun Qiu Bi), and Liu Zhong in "The Battle of Pu Pass" (Zhan Puguan).

做工老生——《徐策跑城》中的徐策

1.3.

靠把老生

靠把老生又称长靠老生。“靠”是京剧的专有名词，就是古代武将所穿的铠甲。身穿铠甲，在京剧里叫作披靠或扎靠。“把”是“把子”的简称，就是兵器，俗称“刀枪把子”。凡是身披铠甲、手持兵器、擅长武功的老生角色，都被称作靠把老生。例如《定军山》《阳平关》中的黄忠，《失街亭》中的王平，《战太平》中的花云，《八大锤》中的岳飞，《武昭关》《战郢城》中的伍子胥，等等。

Kaoba Lao Sheng

Kaoba Lao Sheng (aged male roles in armour bearing weapons), is also known as Changkao Lao Sheng (aged male roles in armour with four flags attached to their back). "Kao" is a special term in Peking Opera, referring to the armour worn by ancient military officers. In Peking Opera performances, wearing armour is known as Pi Kao (lit. "to wear armour"), or Zha Kao (lit. "to tie the armour on" – the costume consists of many separate elements that must be tied on to the actor). The term "Ba", an abbreviation for Bazi, which itself is an abbreviation of Dao Qiang Bazi (swords and spears), refers to the weapons that military characters carry. For this reason, the aged male roles that wear armour, hold weapons and are good at martial arts are called Kaoba Lao Sheng. These include, Huang Zhong in "Battle at Dingjun Mountain" (Dingjun Shan), "Yangping Pass" (Yanping Guan), Wang Ping in"Losing the Battle in Jieting" (Shi Jieting), Hua Yun in "Battle in Taiping City" (Zhan Taiping), Yue Fei in "Taking Turns to Fight against the Four Pairs of Mallets" (Ba Da Chui), Wu Zixu in "Wuzhao Pass" (Wuzhao Guan), and "Battle in Ying City" (Zhan Yingcheng).

箭衣老生

1.4.

在京剧中，穿着箭衣的武老生被称为箭衣老生。箭衣是扮演帝王、驸马及高级武官所穿的常服。箭衣老生在戏里不一定都有武打，但所扮演的角色大多数是武人出身。如《武家坡》中的薛平贵，《南阳关》中的伍云召，《打登州》中的秦琼等。

Jianyi Lao Sheng

Jianyi Lao Sheng are aged male roles who wear an archer's uniform. Jianyi (the archer's uniform) is a kind of military uniform worn by the Emperor, the Emperor's son-in-law, or by Senior Military Officers. Most of the Jianyi Lao Sheng roles have a military background, though they do not necessarily fight in the performance. Examples include Xue Pinggui in "Wujia Slope" (Wujia Po), Wu Yunzhao in "Nanyang Pass" (Nanyang Guan), Qin Qiong in "Attack in Dengzhou City" (Da Dengzhou).

◎ 著名演员（一）

谭鑫培（1847 ~ 1917）：卓越的京剧艺术大师，他博采众长化为己有，开创京剧史上第一个老生流派——谭派，成为京剧界有史以来传人最多、流布最广、影响最大的老生流派。行内有“无腔不学谭”之说。他与汪桂芬、孙菊仙被誉为“新三鼎甲”。除演戏外，谭鑫培还整理加工剧目，修订剧本中戏词不通、悖理或冗沓处。其主演的《定军山》被拍摄成电影，成为中国第一部电影。

表演风格：谭鑫培嗓音甜亮，略有沙音，被称为“云遮月”。行腔高音刚而不直不硬，低声柔而不沉不懈，嘎调宽亮且有很高的润腔技巧。谭鑫培善于突破，改变了以往老生唱

箭衣老生——《武家坡》中的薛平贵

腔全凭嗓子和气力的直腔直调、高音大嗓的唱法，将青衣、老旦、花脸的唱腔及昆曲、京韵大鼓、单弦牌子曲等唱法融会贯通，融入老生唱腔中，创造出细腻婉转的演唱风格，因而他的演唱玲珑活泼，变化多端，于平淡中见灵巧，使唱念更生活化，富有感染力，增强了老生唱腔的艺术表现力。

谭鑫培戏路宽泛，安工、衰派，靠把老生都很擅长。其代表剧目有《定军山》《阳平关》《战长沙》《捉放曹》《洪羊洞》《四郎探母》《李陵碑》《游龙戏凤》《击鼓骂曹》《乌龙院》《打渔杀家》《空城计》等。

Famous Artist（Ⅰ）

Tan Xinpei (1847 ~ 1917): As an excellent artistic master of Peking Opera, Tan Xinpei adopted the best skills from many other performers and incorporated them into his own performance, and founded the first artistic school of Lao Sheng – "Tan" - in the history of Peking Opera. The Tan school is the most widespread school Lao Sheng, with the greatest influence and most disciples. There is a popular saying in the profession: "there is no tune that is not learned from Tan". Tan Xinpei was universally acknowledged as one of "The Best New Three" of Lao Sheng roles together with Wang Guifen and Sun Juxian (the Best Old Three Lao Sheng roles were Cheng Changgeng, Yu Sansheng, and Zhang Erkui). Besides Peking Opera performance, Tan Xinpei had sorted and processed many Peking Operas and revised scripts to amend incorrect words or unrealistic plots and remove extraneous detail. The Peking Opera "Battle at Dingjun Mountain" (Dingjun Shan), in which Tan had starred, was made into the first film in China.

Performance style: Tan' s voice was sweet and slightly husky, and was compared to "the moon covered by a cloud". When singing, his vibrato was firm but not rigid, his voice was soft and bright, and supported by strong technique. He could make bright and broad "Ga" tunes (extremely high pitch for a particular word in Peking Opera) with great skill. Tan devoted himself to the innovation of Peking Opera performance, improved the previous singing method of Lao Sheng in which the tunes were straight and high, supported only by the strength of performer' s throat and breath. Tan referred to the singing techniques of Kunqu Opera, as well as Jing Yun Da Gu (a form of storytelling in the Beijing dialect with drum accompaniment), and Dan Xian (storytelling accompanied by a Sanxian, a three-stringed plucked instrument), and integrated the tunes of Qingyi (young female roles in Peking Opera), Lao Dan (old female roles in Peking Opera), and Hualian (male roles with facial painting) into the tunes of Lao Sheng. Tan created a delicate and melodious style of singing in which his performance was lively and flexible, found skill in the mundane, and was full of life. Tan' s innovation greatly enhanced the artistic capabilities of the Lao Sheng singing voice.

Tan Xinpei was good at performing Angong Lao Sheng (sedate aged male roles), Shuaipai Lao Sheng (aged male roles in decline) and Kaoba Lao Sheng (aged male roles wearing armour), and his masterpieces included: "Battle at Ding Jun Mountain" (Dingjun Shan), "Yangping Pass" (Yanping Guan), "Battle in Changsha City" (Zhan Changsha), "Capture and Release of Cao Cao" (Zhuo Fang Cao), "Hongyang Cave" (Hongyang Dong), "Si Lang Visits His Mother" (Silang Tan Mu), "The Tombstone of Li Ling" (Li Ling Bei), "The Playful Dragon Teases The Phoenix" (You Long Xi Feng), "Scolding Cao Cao Beating a Drum" (Ji Gu Ma Cao), "Black Dragon Courtyard" (Wu Long Yuan), "The Fisherman' s Revenge" (Da Yu Sha Jia), and "Empty-Town Ruse" (Kong Cheng Ji).

◎ 著名演员（二）

余叔岩 (1890 ~ 1943)：全面继承了祖父余三胜（“前三鼎甲”，习称“老生前三杰”之一）、老师谭鑫培（“新三鼎甲”，习称“老生后三杰”之一）丰厚的艺术传统，特别深入地钻研了集前辈技艺精华之大成的谭派艺术。凭着他的文化素养，在全方位掌握谭派特点和规律的基础上，对于所演剧目的词句、辙韵多有润色，使之顺畅合理，又从博雅精深处寻求发展，达到了“出蓝胜蓝”的再创造境界，创立了自己的艺术流派，世称“余派”，成为老生发展史上里程碑式的人物，对后世老生艺术的发展有深远的影响。与言菊朋、高庆奎、马连良并称“前四大须生”。

表演风格：余派的主要特色为唱腔韵味醇厚，刚柔并蓄，演唱技巧丰富，台风端庄大方。其嗓音甘甜清醇，精于发音技巧，又结合气息运用技巧，做到高音清越，低音苍劲，立音峭拔雄浑，擞音圆润，颤音摇曳多姿，最大限度地获得用嗓的自由。余派唱腔以情带腔，旋律多变，抑扬顿挫，声断意连，运腔技巧要求极严，较谭（鑫培）派更为精巧雅致。余叔岩的念白注重音韵，字音准确清晰。讲究吐字发声，对于“三级韵”规律运用纯熟，善用虚词，起落清楚，反切极好。在做功方面，余派表演在继承谭鑫培的基础上又有较大变化，身段讲究戏理剧情，贴近生活实际，不事花哨，生动妥帖，着重于表现人物的内心活动。余叔岩武功根底厚实，靠把戏的开打、亮相等功夫独到。

其代表剧目有《乌盆记》《伍子胥》《定军山》《阳平关》《李陵碑》《游龙戏凤》《打渔杀家》《空城计》《四郎探母》《清风亭》《珠帘寨》等。《搜孤救孤》为余叔岩创演的名剧，是余派的传世之作。

Famous Artist（Ⅱ）

Yu Shuyan (1890 ~ 1943): Inheriting the rich skill of his Grandfather, Yu Sansheng (one of "The Best Old Three" Lao Sheng roles) and his master Tan Xinpei (one of "The Best New Three" Lao Sheng roles), Yu Shuyan delved most deeply into the performing art of the Tan School, which had integrated the essence of Lao Sheng performing skill of its predecessors into its own artistic style. Based on his high cultural accomplishment, Yu Shuyan comprehensively mastered the characteristics and rules of Tan's artistry, polished the words, sentences and rhyme schemes of the scripts he was performing to make them more logical. Yu developed Tan' s art into a more accomplished and profound new artistic school of Lao Sheng performance, known as "Yu". Yu Shuyan became a landmark character in the historical development of the performing art of Lao Sheng, and had a far-reaching influence on its later development. Yu Shuyan was acknowledged as one of "The Former Four Great Lao Sheng", together with Yan Jupeng, Gao Qingkui, and Ma Lianliang.

Performance style: The main artistic features of the Yu (Shuyan) School included the melodious singing of tunes in a manner that was both firm and mellow, rich singing skills and a dignified style on stage. Yu' s voice was sweet and clear. By practicing skills of pronunciation and breath control, Yu made his high-pitched voice clear and penetrating, his low-pitched voice strong and vigorous, his li voice (a traditional singing technique in

Peking Opera, where the voice is produced by resonation through cavities in the head) powerful and steady, his sou voice (another traditional singing technique in Peking Opera, where the voice is centred in the larynx) round and mellow, and his vibrato swayed and glittered. Yu used his voice freely, and his singing was famous for its touching tunes, fluctuating melody, measured cadences, lingering charm when the singing stopped, and the dextrous and proper use of skills, which was more elegant and refined than Tan Xinpei's. For the spoken parts, Yu Shuyan emphasized accurate pronunciation and enunciation, skilfully practised the "Three-Level Rhyme", was proficient in using function words, paid close attention to the clarity of rise and fall when intoning words, with excellent Fanqie method (a traditional method of indicating the pronunciation of a Chinese character by using two other Chinese characters, the first having the same consonant as the given character and the second having the same vowel and tone). In terms of performance, the Yu School greatly modified the techniques inherited from Tan Xinpei's performance. Yu's postures were vivid and correctly executed, but not showy; they were closely connected to daily life, and appropriate to the plot of the play. His movements focused on expressing the psychological state of characters. Yu Shuyan was proficient in the basic skills of martial arts, especially in performing fight routines, and striking a memorable pose (a pause in the choreography where the actor demonstrates their presence, known as 'liang xiang') on the stage.

Yu's representative masterpieces included: "Redressing a Grievance" (Wu Pen Ji), "Wu Zixu" (Wu Zixu), "Battle at Dingjun Mountain" (Dingjun Shan), "Yangping Pass" (Yangping Guan), "The Tombstone of Li Ling" (Li Ling Bei), "The Playful Dragon Teasing The Phoenix" (You Long Xi Feng), "The Fisherman's Revenge" (Da Yu Sha Jia), "Empty-Town Ruse" (Kong Cheng Ji), "Si Lang Visits His Mother" (Silang Tan Mu), "Qingfeng Pavilion" (Qingfeng Ting), and "Zhulian Village" (Zhulian Zhai). "Searching And Rescuing The Orphan" (Sou Gu Jiu Gu) was a famous Peking Opera which was created by, and starred, Yu Shuyan, and has become the enduring work of the Yu School.

◎ 著名演员（三）

周信芳（1895 ～ 1975）：京剧老生表演艺术家，艺名“麒麟童”，京剧老生“麒派”艺术创始人。周信芳勇于创造，在继承的基础上多了大量的改革创新，除对传统剧目做去芜存菁的整理改动之外，无论唱、念、做、打与剧目、唱词、服装、扮相等方面均有适合于自己风格而与众不同的设计。

表演风格：周信芳的唱腔以苍凉遒劲为特色，朴而不直，顿挫有力，往往有极富曲折跌宕之处，尤其注意抒发人物感情，高拨子、汉调等唱腔有独特的韵味。麒派念白江浙方音重，苍劲、饱满，富于力度，犀利老辣，咬字清晰，富有音乐性，且善用语气词，有时接近于口语，生活气息浓厚，无论表达风趣、庄重、愤恨、哀伤的情绪，语气都极为自然生动。在做功方面，麒派做功真切生动，强调唱、念、做、打的综合统一性，在表演中运用水袖、身段、步法，结合眼神和面部表情，吻合剧情的特定环境及人物思想，达到浓墨重彩的效果。

周信芳的代表剧目有《徐策跑城》《乌龙院》《萧何月下追韩信》《四进士》《二堂舍子》《董小宛》等。

Famous Artist（Ⅲ）

Zhou Xinfang (1895 ～ 1975): Zhou Xinfang, whose stage name was Qi Lintong, was a great Peking Opera performing artist, and the founder of the Qi School of Lao Sheng. Drawing upon the techniques he inherited from his predecessors, Zhou Xinfang created new innovations and reformed the performance art of the Lao Sheng role. In addition to making changes to the repertoire on the basis of quality, Zhou Xinfang modified the skills of singing, recitation, dance-acting, and fighting, revised the lyrics of scripts, and redesigned costumes and facial makeup designs to be suitable for his own performing style, making it unique.

Performance style: The performance of arias by Zhou Xinfang was characterized by a cool sturdiness. It was simple and straightforward, but powerful. The melody was full of twists and turns, paying particular attention to the expression of emotion. Zhou added Gaobozi (a type of tune that originates from Hui opera) and Handiao (generally called Erhuang, a type of tunes of traditional Chinese opera, popular in Hubei and Shan'xi Province) into his singing, which gave it a distinctive charm. The spoken parts of the Qi School had an obvious accent from Jiangsu and Zhejiang province, which sounded vigorous, firm, strong, sharp and skilful. With clear and musical pronunciation, and the flexible application of modal particles, Zhou's speaking was close to the language used everyday. It was full of the rich flavour of real life, and had a natural and vivid tone when expressing humour, solemnity, hatred, and sorrow. He naturally and vividly conveyed emotion. In terms of acting technique, the Qi School emphasized the total integration of singing, recitation, dance-acting, and fighting, and devoted particular care to the use of long sleeves, postures, and footwork, which when combined with eyes and facial expressions, matched the plot and the particular emotion or situation of the character. These had a very intensive effect in performance.

The best-known masterpieces of Zhou Xinfang include: "Xu Ce Runs on the City Wall" (Xu Ce Pao Cheng), "Black Dragon Courtyard" (Wu Long Yuan), "Xiao He Chases After Han Xin in the Moonlight" (Xiao He Yue Xia Zhui Han Xin), "Four Successful Candidates of the Imperial Examination" (Si Jinshi), "Saving Chen Xiang" (Er Tang She Zi), and "Dong Xiaowan" (Dong Xiaowan).

◎ 著名演员（四）

杨宝森（1909 ~ 1958）：京剧老生演员，与马连良、谭富英、奚啸伯并称为“四大须生”，杨派艺术的创始人。8 岁学艺，后带艺搭班入斌庆社科班。14 岁登台。他专心研习余叔岩的表演艺术，创造性地继承发展了谭派和余派艺术，唱法、行腔自成一派，世称“杨派”。“杨派”成为当今流传最广、影响最大的京剧流派之一，更造就了“十生九杨”的京剧生行格局。杨宝森虽以唱功为主，但是做派也有相当的功力。

表演风格：杨宝森的唱腔、唱法，纯以韵味取胜，他的嗓子宽厚而低沉，唱腔简洁大方，颇具余派稳健含蓄、韵味醇厚的特点，也体现了余派刚柔相济、圆而不滑、平中见奇的艺术特色，虽缺少大幅度的起伏跌宕，却于细微处体现丰富的旋律，细腻而不琐碎。比如，杨宝森在《空城计》中老生的大段唱功，于心平气和中蕴含深厚的感染力，唱腔古朴恬淡，婉转深沉，悲凉凝重，以表现人物冷静、机智的性格及胸有成竹、运筹帷幄的心理特点。

杨宝森代表剧目有《伍子胥》《失 · 空 · 斩》《击鼓骂曹》《搜孤救孤》等。

Famous Artist（Ⅳ）

Yang Baosen (1909 ~ 1958): Yang Baosen, a great Lao Sheng performing artist, and the founder of Yang School of Lao Sheng. He was acknowledged as one of the "The Later Four Great Lao Sheng" together with Ma Lianliang, Tan Fuying, and Xi Xiaobo. Yang was apprenticed at the age of eight, and later joined the Bin Qing She (an old famous Peking Opera school and troupe in Beijing) appearing on the stage for the first time aged fourteen. Yang concentrated on Yu Shuyan's performing arts, learnt and then developed the artistic style of the Tan (Xinpei) School and the Yu (Shuyan) School, and established a new singing technique, which became known as the Yang (Baosen) School. The Yang School is one of the most widespread and influential Peking Opera schools, so much so that there is a maxim that states there are "Nine Yang out of every Ten Sheng" roles on stage. Yang focused on singing, though his acting was also considered skilful.

Performance style: Yang won his success by his singing technique, and his command of arias. His voice was deep and broad, and his singing was simple and elegant. His voice was characterized by a restraint, subtlety and mellowness, which is the embodiment of the Yu school. Yang's singing was full of melodic detail, though without great undulation, and remained delicate without being trivial. This encapsulates Yu Shuyan's artistic style of balancing vigour and suppleness. For instance, in a protracted section of singing in the play the "Empty-Town Ruse" (Kong Cheng Ji), Yang's singing impressed the calmness of the character upon the audience, through a combination of simple and tranquil, mellow and deep, bleak and dignified melodies. This expressed the disposition of Zhuge Liang; his calmness and wisdom, as well as his self-belief and ability to strategize.

Yang Baosen's representative masterpieces include: "Wu Zixu" (Wu Zixu), "Losing the Battle in Jieting" (Shi Jieting), "Empty-Town Ruse" (Kong Cheng Ji), "Executing Ma

Su" (Zhan Ma Su), "Beating a Drum to Curse Cao" (Ji Gu Ma Cao), and "Searching And Rescuing The Orphan" (Sou Gu Jiu Gu).

◎ 著名演员（五）

马连良（1901 ~ 1966）：回名尤素福，原籍陕西扶风，生于北京。中国著名京剧艺术家，老生行当的代表性人物之一，“马派”艺术创始人，京剧“四大须生”之首。马连良对艺术学而不厌，虚怀若谷，为京剧艺术事业的精益求精贡献一生。

表演风格：马派念白和马派唱腔互为表里，彼此依托，统一于马派声腔艺术之中。马连良在经常演出的剧目中，念白有时老辣，有时苍劲，有时是幽默风趣的声调，有时是忠告谏劝的语气，用以表达人物的不同性格和感情需要。他善于将念白处理得像唱一样具有强烈的艺术感染力，悦耳动听。他的表演将手、眼、身、步合为一体，准确严谨，凝重潇洒，将人物蕴含的复杂感情，节奏鲜明地形之于外，形成独具艺术魅力的马派艺术。

马连良代表的剧目有《春秋笔》《临潼山》《清官册》《一捧雪》《四进士》《乌龙院》《桑园会》《打登州》《南阳关》《苏武牧羊》等。

Famous Artist（V）

Ma Lianliang (1901 ~ 1966): also named You Sufu, was originally from Fufeng, Shanxi Province, was born in Beijing. A famous representative of Lao Sheng roles and the founder of the Ma School of Lao Sheng. Ma Lianliang is acknowledged as the chief of "The Later Four Great Lao Sheng". Ma had an insatiable desire and modest attitude to learn about art, and devoted himself to constantly developing and improving the performance artistry of Peking Opera.

Performance style: The spoken and singing parts of the Ma School are mutually dependent, relying on each other to achieve unification in the sonic artistry of the Ma School. In Ma's oft-performed repertoire, the spoken parts are sometimes sophisticated, vigorous or humorous, sometimes admonishing or persuasive, according to the personality and emotion of the character. Ma was skilful in interpreting recited lines with charming melodic-like intonation. Ma's acting, combined with the movements of his hands, expression of his eyes, posture and footwork, was integrated, accurate, precise, dignified, and elegant, which vividly displayed the complicated emotions of the characters and formed the unique charm of the artistic style of the Ma School.

Ma Lianliang's classical repertoire included: "The Honest Pen" (Chun Qiu Bi), "Lintong Mountain" (Lintong Shan), "List of Upright Officials" (Qingguan Ce), "A Handful of Snow" (Yi Peng Xue), "Four Successful Candidates of the Imperial Examination" (Si Jin Shi), "Black Dragon Courtyard" (Wu Long Yuan), "The Reunion in the Mulberry Field" (Sang Yuan Hui), "Attack in Dengzhou City" (Da Dengzhou), "Nanyang Pass" (Nanyang Guan), and "Su Wu Herding Sheep" (Su Wu Mu Yang).

1.5.

红生

红生是指面部化妆，用红色勾画脸谱的老生。主要扮演关羽和赵匡胤这两个人物。红生在京剧行当中独树一帜，它既不同于花脸，也不同于武生和老生，却需有花脸的功架、武生的功底，以及老生高亢醇厚的念白和唱法。红生扮相威风凛凛，器宇轩昂，具有威严震慑之美感。代表剧目有《古城会》《华容道》《汉津口》《战长沙》《水淹七军》《走麦城》《龙虎斗》等。

Hong Sheng

Hong Sheng (red-faced male roles) refers to a type of Lao Sheng (aged male roles) that wears red facial make-up. The main characters of this role are Guan Yu (a famous general of the Three Kingdoms periods),

红生——《走麦城》中的关羽

and Zhao Kuangyin (the founding emperor of Song Dynasty). As a role with a unique style of its own, the Hong Sheng is different from other Hualian (also known as Jing, painted face male roles), Wu Sheng (martial male roles) and other types of Lao Sheng (aged male roles). A Hong Sheng actor has to integrate the movements and postures of the Hualian, the basic martial training of the Wu Sheng role, and the resounding and mellow speaking and singing of voice of the Lao Sheng role into his own performance. The appearance of the Hong Sheng role in costume and makeup is always majestic and dignified, with an awe-inspiring and imposing air.

The representative repertoire includes: "Reunion at Gucheng Town" (Gucheng Hui), "Huarong Road" (Huarong Dao), "Hanjin Estuary" (Hanjin Kou), "The Battle in Changsha City" (Zhan Changsha), "Drowning the Seven Armies" (Shui Yan Qi Jun), "Defeat in Maicheng Town" (Zou Maicheng), and "The Fight between the Dragon and the Tiger" (Long Hu Dou).

◎ 著名演员

王鸿寿（1850 ~ 1925）：艺名三麻子。他的表演风格形成于清末，是京剧红生行当的创始人，所表演的红生戏别具一格，在继承前辈艺人米喜子、程长庚等表演艺术的基础上，有很大的创造与发展，被尊为“红生之祖”。

表演风格：王鸿寿擅演关羽，他突破了原北派（即米、程一系）关羽戏的局限，变北派塑造关羽形象以肃穆庄严为主的演法为南派的威武激昂，冲淡对关羽的过分“神化”，而加强其刚劲雄武的描绘。王鸿寿对关羽的扮相、服饰、功架、武打、唱、念、表演做了多方面的丰富与创新：变关羽的揉胭脂脸为勾银朱脸，增加了脸谱的线条，加重脸谱的色彩；变关羽戴黑满或黑五绺髯口为黑三绺；为关羽设计了独用的夫子盔、掩心甲、软靠，特制了青龙刀；借鉴关羽的各种塑像、画像，设计了多种优美的造型，又设计了特殊的趟马和刀花亮相；在唱法上，他以高亢的嗓音唱徽调，并以昆曲吹腔、梅花板、南韵拨子、西皮、二黄、唢呐二黄等互相配用的方法作为关羽独有的唱腔；在做派上，融合文武老生、武生及架子花脸的表演特色，使之兼有庄重、沉雄、勇猛、威武等多种特色，并增设一名马童来丰富关羽的舞蹈动作。

王鸿寿擅演的红生戏有《单刀会》《水淹七军》《温酒斩华雄》《白马坡》《斩颜良》《诛文丑》《灞桥挑袍》《华容道》《古城会》《战长沙》等。

Famous Artist

Wang Hongshou (1850 ~ 1925): The performance style of Wang Hongshou, with the stage name of San Mazi, was formed in the late Qing dynasty. He was the founder of the Hong Sheng role. The performance of Hong Sheng drama is unique. Though it was developed from other artists - Mi Xizi and Cheng Changgeng - Wang initiated the unique artistry of the Hong Sheng role, and he is now regarded as the progenitor of the Hong Sheng role.

Performance style: Wang specialized in playing Guan Yu. He transcended the limitations of performing the character in the Northern style, i.e. in Mi and Cheng's style, and changed the image of Guan Yu from "northern solemnness" to "southern mightiness and passion". This weakened the excessive deification of Guan Yu, and strengthened the depiction of the character to focus on his power and heroism. Wang Hongshou created a number of new aspects for the performance of Guan Yu, such as his facial makeup, costume, postures, fighting patterns, dance-acting, singing and recitation. Wang increased the lines and intensified the colour of the face painting by altering the method of applying rouge to the cheeks by fingers, and by delineating the outline of the facial features and wrinkles with pen. He changed the original rankou (artificial beard) from a heiman (full black beard) into heisan (black beard consisting of three separate strands), designed the warrior's helmet, soft armour with breastplate, and a special Qing Long falchion (shaped like a crescent moon, and decorated with a carved dragon) exclusively for Guan Yu. After consulting statues and portraits of Guan Yu, Wang created various postures and movements for Guan Yu, particularly with regard to riding a horse and manipulating the falchion. In terms of singing technique, Wang adopted Huidiao (one of traditional local operas in Anhui province) tunes and combined them with Chuiqiang (a type of traditional opera tunes, accompanied with bamboo flute) of Kunqu (one of the most ancient traditional operas in China, originating from the end of Yuan Dynasty more than six hundred years ago in southern Jiangsu), Meihuaban (a type of rhythmic form in traditional opera music), Nanyun Bozi (a type of tune in Huidiao with a distinctly southern style), Xipi (one of the chief modes in Peking Opera, roughly equivalent to major in Western music), Erhuang (one of the chief modes in Peking Opera, roughly equivalent to minor in Western music), and Suona Erhuang (Erhuang tune accompanied with Suona, a traditional Chinese reed instrument) as unique tunes to be used exclusively for Guan Yu. For acting, Wang Hongshou integrated various acting styles of civil (wen) and military (wu) roles, including Lao Sheng, Wu Sheng (martial male roles), and Jiazi Hualian (male roles with painted face, that emphasize acting skills), into the performance of the Hong Sheng. The result was a performance that combined solemnity, calmness, bravery and mightiness. Wang also created a new character (a Groom) for Guan Yu to enrich the character's dance-acting.

The classic plays of the Hong Sheng role that Wang Hongshou starred in included: "Attending the Banquet Alone" (Dan Dao Hui), "Drowning the Seven Armies" (Shui Yan Qi Jun), "Chopping Hua Xiong Hardly Waiting for the Liquor Cooling" (Wen Jiu Zhan Hua Xiong), "Battle at Baima Slope" (Baima Po), "Chopping Yan Liang" (Zhan Yan Liang), "Killing Wen Chou" (Zhu Wen Chou), "Lifting up the Given Robe with Falchion at Baling Bridge" (Baqiao Tiao Pao), "Huarong Road" (Huarong Dao), "Reunion at Gucheng Town" (Gucheng Hui), and "The Battle in Changsha City" (Zhan Changsha).

红生——《华容道》中的关羽

小生

2.

小生为京剧中生行的一种，扮演青年男性角色，特点是不戴髯口，扮相清秀英俊，衣服颜色鲜艳明快。在表演上最大的特点是唱和念以假嗓为主，兼用真嗓，多念韵白。小生的假嗓是龙、凤、虎音结合，挺拔高亢，比真嗓尖、细、高，以表现年少气盛，这样就从声音上与老生有所区别。小生的假嗓发音更加清脆、刚健、宽亮，又区别于旦角发音的柔媚、娇艳。小生的动作造型的基调是风流儒雅、俊逸飞动。

小生主要分为文小生和武小生。有人认为娃娃生也是小生的一种。小生流派有姜（妙香）派、叶（盛兰）派等。

Xiao Sheng

Xiao Sheng (young male roles) are one of main types of Sheng (male) roles in Peking Opera, usually portraying young male characters who have delicate features, a handsome appearance and wear brightly-coloured costumes. In terms of facial make-up, the Xiao Sheng do not wear rankou (artificial beards). The most significant feature of the Xiao Sheng role is the use of a falsetto voice when singing and speaking, occasionally mixed with true voice and the adoption of yunbai (also known as zhongzhouyun, a form of stylized speech which offers the character a sense of decorousness) when speaking. The falsetto of Xiao Sheng is the mixture of "dragon sound" (a powerful high pitched sound, so called because it supposed to sound like a dragon howling), "tiger sound" (a deep and low pitch, supposed to sound like a tiger roaring), and "phoenix sound" (a clear and resonant sound, likened to the imagined song of a phoenix singing), which is sharper, thinner and higher than the sound of true voice. This shows the spirit and energy of a young man, thus aurally distinguishing the role from the Lao Sheng (aged male roles). In comparison with the falsetto used by the Dan

(Female roles), the falsetto of the Xiao Sheng is clearer, wider, brighter and has more power (the Dan voice is considered comparatively soft and gentle). The postures and movements of Xiao Sheng on the stage are elegant and attractive.

There are two main types of Xiao Sheng roles: Wen Xiao Sheng (civilian young male roles) and Wu Xiao Sheng (military young male roles). Some also classify Wawa Sheng (child roles) as Xiao Sheng roles. The important schools of Xiao Sheng include: Jiang (Miaoxiang), Ye (Shenglan).

文小生

2.1.

文小生多扮演气度雍容、儒雅，充满书卷气的青年男性角色。从表演技术和人物类型上可细分为：袍带小生、巾生、穷生等。

Wen Xiao Sheng

Wen Xiao Sheng (civilian young male roles) mainly portray the graceful, refined and scholar-like young male roles. According to the demands of each character and the performance skills they require, the role can be further classified into Paodai Xiao Sheng (young male roles who wear a court robe), Jin Sheng (young male roles wearing a civilian kerchief), and Qiong Sheng (poor young male roles).

文小生——《三堂会审》中的王金龙

袍带小生

2.1.1.

袍带小生又名袍带生、官生、冠生、纱帽小生。扮演做官的青年人，大部分为文人形象，戴纱帽、穿蟒袍，风度雍容端庄，儒雅稳重，如《玉堂春》中的王金龙、《陈三两爬堂》中的李凤鸣等。还有一种是穿官衣的袍带小生，表示官阶比较低，如《奇双会》中的赵宠、《御碑亭》中的柳生春（考中进士以后）、《清风亭》最后出场的张继保等。

Paodai Xiao Sheng

Paodai Xiao Sheng (young male roles who wear a court robe), is also known as Guan Sheng (young male official roles), Guan Sheng (young male roles wearing an official's hat), and Shamao Xiao Sheng (young male roles wearing a gauze cap). They mainly portray young members of the literati and carry a dignified, graceful, and poised deportment. They wear a shamao (gauze cap) and a mangpao (court robe). Examples include Wang Jinlong in "Yu Tangchun" (Yu Tangchun), Li Fengming in "Chen Sanliang Standing Trial" (Chen Sanliang Pa Tang). Sometimes Paodai Xiao Sheng wear Guanyi (official robes) to portray low ranked officials, such as Zhao Chong in "Li Qi Reuniting with His Son And Daughter" (Li Qi Hui), Liu Shengchun (after the imperial examinations) in "The Pavilion of the Imperial Stele" (Yubei Ting), and Zhang Jibao in the last scene of "Qingfeng Pavilion" (Qingfeng Ting).

袍带小生——《陈三两爬堂》中的李凤鸣

2.1.2.

巾生

巾生又名褶子生或扇子生。一般为头戴文生巾，手执扇子，身穿绣花褶子的风流潇洒、文质彬彬的青年男子。在京剧舞台上，拿扇子不是为了凉快，而是因为扇子本身就是一种舞蹈道具，用来辅助表现角色风流倜傥、温文尔雅的气度，并配合演员的舞蹈动作。巾生主要出现在爱情戏中，故常与旦角配戏。如《白蛇传》中的许仙、《拾玉镯》中的傅朋、《西厢记》中的张生、《人面桃花》中的崔护等。

Jin Sheng

Jin Sheng (young male roles who wear a civilian kerchief), is also known as Xuezi Sheng (young male roles who wear a casual coat), or Shanzi Sheng (young male roles who hold a fan). The wearing of these garments indicates to the audience that this role type is graceful, gentle and refined in manner. On the stage, characters do not hold a folding fan because they are hot and need to cool themselves, but to demonstrate their elegant and cultured bearing. It is also used by the performer when they move and dance. Jin Sheng mainly appear in romantic stories, and thus usually appear alongside Dan (Female) roles. Examples include Xu Xian in "The White Snake" (Baishe Zhuan), Fu Peng in "Picking Up the Jade Bracelet" (Shi Yuzhuo), Zhang Sheng in "The Romance of the Western Chamber" (Xixiang Ji), and Cui Hu in "Peach Blossom-like Face" (Ren Mian Tao Hua).

巾生——《西厢记》中的张生

2.1.3.

穷生

穷生多扮演穷困落魄的读书人。这些角色一般穿富贵衣。富贵衣是一个专用名词，就是在青道袍上补许多各色补丁，意思是说这些人眼前虽然很穷，但将来仍会飞黄腾达，因此眼下穿的补丁衣服被称为“富贵衣”。穷生在表演上特别注重做功，走路摇摆做态，略带移步，两手常抱胸，以表现人物失意、颓丧、酸腐之状，如《金玉奴》中的莫稽、《状元谱》中的陈大官、《连升店》中的王明芳等。

Qiong Sheng

Qiong Sheng (poor young male roles), mainly play poor intellectuals who have lost their social status. This type of role usually wears a fuguiyi (which ironically translates as "rich and expensive clothes"). Fuguiyi is a black robe adorned with patches of different colours and shapes, stylistically representing the clothes worn by the poor. Nevertheless, the fuguiyi suggests that the character wearing this stylistically tatty garment has a promising future, and will grow to become influential in society. Qiong Sheng roles particularly emphasize acting technique, walking in a swaying manner with a slightly rolling gait, and with arms folded across the chest to show their frustrated and dispirited state, as well as their pedantic temperament. Examples include the character of Mo Ji in "The Story of Soybean Milk" (Jin Yu Nu), Chen Daguan in "The Story of a Number One Scholar" (Zhuangyuan Pu) and Wang Mingfang in "Liansheng Inn" (Liansheng Dian).

穷生——《状元谱》中的陈大官

武小生

2.2.

武小生又称翎子生或雉尾生，因演员头戴插有翎子的盔帽而得名。翎子，也叫雉尾，一米四五长，作为头部的装饰品插在盔帽上。翎子生多扮演扮相英俊，雄健勇武，血气方刚，文武兼备的青年将领。演翎子生一般要有一些武功，主要凭功架，并且用翎子做出许多舞蹈动作。比如《群英会》中的周瑜、《凤仪亭》中的吕布、《穆柯寨》中的杨宗保等，都属于翎子生范畴。

Wu Xiao Sheng

Wu Xiao Sheng (military young male roles), is also known as Lingzi Sheng or Zhiwei Sheng. These alternative names come from the fact that the character often wears a warrior's helmet decorated with lingzi or zhiwei - long pheasant tail feathers that are between 1.4 and 1.5 metres long and are inserted into the helmet as decoration. Lingzi Sheng usually play handsome, valiant, heroic, red-blooded young generals, who are both well-educated and highly skilled in martial arts. Linzi Sheng are generally good at acrobatic fighting and movements that involve manipulating the lingzi feathers. Zhou Yu in "A Meeting of Heroes" (Qun Ying Hui), Lü Bu in "Fengyi Pavilion" (Fengyi Ting), and Yang Zongbao in "Muke Village" (Mu Ke Zhai) are all characters played by this role.

武小生——《群英会》中的周瑜

◎ 著名演员（一）

姜妙香（1890 ~ 1972）：著名京剧表演艺术家，京剧小生流派“姜派”创始人。幼时师从名家学习青衣，9 岁开始唱戏，两年后被戏班接纳为正式演员，开始舞台生涯。姜妙香在 21 岁时因病不得不改学小生。他边学戏边演出，以惊人的毅力战胜了重重困难，终于成为京剧著名小生演员。后与梅兰芳合作演出《玉堂春》，由于二人的演出珠联璧合，倾倒观众，自此开始了他们二人长达 46 年之久的艺术合作之路。

表演风格：姜妙香博采众长且勇于创新，对小生传统唱腔精雕细刻地进行加工改进，创造了许多优美动听的小生唱腔及多种新颖别致的唱法，形成了新的流派，被称为“姜派”。姜妙香扮相儒雅庄重、真挚自然、唱做俱佳，尤精于唱。他嗓音甜润、气力充沛，在改革小生唱腔的过程中吸收青衣的传统唱腔、唱法，去其脂粉气，并将娃娃腔运用于西皮、二黄板式唱段中，丰富和发展了小生的成套唱段。念白讲究声韵和吐字，强调语气的轻重缓急。在做功方面，姜派小生在表演上做式精到利落、技巧纯熟，对人物揣摩得十分细致，能够贴切表达人物的情感，从而深受观众喜爱，流传不衰。姜妙香戏路较宽，擅长袍带生、扇子生、穷生及翎子生。其代表作品有《玉堂春》《状元谱》《四郎探母》《凤还巢》《御碑亭》《连升店》《穆柯寨》《贩马记》等。

Famous Artist（Ⅰ）

Jiang Miaoxiang (1890 ~ 1972): Jiang Miaoxiang was a well-known Xiao Sheng actor, and the founder of the "Jiang" School of Xiao Sheng. When he was young, he learnt the Qingyi (one type of female role in Peking Opera), appeared on stage at the age of nine, and was accepted as a formal actor by a troupe two years later. His career began at this point. By the age of twenty-one, Jiang was forced to change his role-type and began to learn Xiao Sheng. He gave performances whilst still learning the role, and with amazing perseverance overcame the many difficulties to finally became a famous Xiao Sheng actor. Later, Jiang co-starred with Mei Lanfang in "Yu Tangchun" (Yu Tangchun), and audiences fell in love with what seemed a perfect pairing. Jiang and Mei collaborated successfully for forty-six years.

Performance style: Jiang Miaoxiang was a courageous pioneer over a long period.

He refined the melodies of the Xiao Sheng role and created many beautiful new ones, as well as developing new singing techniques. This led to the formation of a new School of Xiao Sheng performance art, known as the "Jiang" School. Jiang's appearance on stage was elegant and dignified, natural and sincere. He was excellent at both singing and acting, particularly at singing. His voice was sweet and full of energy. He improved Xiao Sheng melodies by absorbing the traditional singing tunes and methods of the Qingyi role he had learnt as a child, stripped the technique of its femininity, and adopted the tunes of the Wawa Sheng in the arias of Xipi and Erhuang (the two chief modes in Peking opera music) to enrich and develop the existing repertoire of Xiao Sheng arias. In terms of recitation skills, Jiang devoted particular care to the pronunciation and enunciation of each word, and stressed the intensity and speed of each tone. When acting, Jiang's movements were neat, precise, and skilful. Jiang skilfully crafted the characters, and was able to describe the emotions of each character accurately. Jiang's art was widely admired by audiences. Jiang Miaoxiang was adept at all the types of Xiao Sheng roles including the Paodai Xiao Sheng (young male roles in court robe, usually officials), Shanzi Sheng (young male roles holding a fan, usually scholars), Qiong Sheng (poor young male roles) and Lingzi Sheng (young male roles wearing warriors helmets decorated with long pheasant tail feathers, usually young generals). Jiang's masterpieces included: "Yu Tangchun" (Yu Tangchun), "The Story of a Number One Scholar" (Zhuangyuan Pu), "Si Lang Visits His Mother" (Silang Tan Mu), "Phoenix Returning To Its Nest" (Feng Huan Chao), "The Pavilion of the Imperial Stele" (Yubei Ting), "Liansheng Inn" (Liansheng Dian), "Muke Village" (Mu Ke Zhai), and "The Experience of the Horse-Trader" (Fanma Ji).

◎ 著名演员（二）

叶盛兰（1914 ~ 1978）：著名京剧表演艺术家，京剧小生流派“叶派”创始人。叶盛兰对程门本派全面继承，并在多年的舞台实践中有重大的突破和发展，青出于蓝而胜于蓝，把程派小生艺术推向又一个新的高峰。叶盛兰性格坚毅，幼年练功学戏特别刻苦，备受赞誉。他扮相英俊洒脱，嗓音清醇宽亮、圆润甜美；表演稳重，极善刻画人物；文武兼备，昆乱不挡，擅演雉尾生、扇子生、袍带生、穷生、武生等各色人物，享有“活周瑜”“活吕布”的美称。

表演风格：叶派小生的唱功刚柔相济，清新健美且华丽遒劲。在唱法上，叶盛兰博取众家精华，并研创了反西皮唱腔，注重展现小生的阳刚之美，并擅以丰富多变的唱腔塑造人物。如饰吕布，他在《辕门射戟》中的西皮唱腔挺拔豪放，盛气凌人，而在《白门楼》中的唱腔则凄婉迂曲、悲怆动人。在做功身段方面，叶盛兰表演细腻，扮武将壮武健爽，英气逼人；演文生清秀飘逸，富有书卷气。同为雉尾生，叶盛兰演的周瑜和吕布有鲜明的不同特色：周瑜需突出“儒”字，而吕布则侧重“蛮”字。演周瑜是颔首、舒眉、抬眼、抿唇、撮口、梗劲迈步，要抬得重，落得稳；而吕布则是扬头、瞪眼、撇唇、张口、松劲晃身，摆臂迈步，要抬得高，落得飘。叶盛兰还总结出水袖程式的四字要诀：“沉、扬、柔、落。”如周瑜要沉，吕布要扬，梁山伯要柔，许仙要落。擅用程式技法来表现人物的性格特征是“叶派”表演艺术的主要特点。其代表作品有《桃花扇》《玉簪记》《凤还巢》《白蛇传》《群英会》《游园惊梦》《周仁献嫂》等。

Famous Artist（Ⅱ）

Ye Shenglan (1914 ~ 1978): was a famous actor and the founder of the "Ye" School of Xiao Sheng. After mastering the "Cheng (Jixian)" School, Ye Shenglan significantly developed and improved the Cheng style during his years of practice on the stage. Ye was highly praised for his determination and hard work in training when he was young. Ye's voice was mellow, bright, and sweet, and he was handsome in appearance, certain in performance, and skilful in portraying characters. With great accomplishment, Ye Shenglan was good at playing various of types of operas and styles of roles, such as Zhiwei Sheng (young male roles wearing warriors helmets decorated with long pheasant tail feathers, usually young generals), Shanzi Sheng (young male roles holding a fan, usually scholars), Paodai Sheng (young male roles in court robe, usually officials), and Qiong Sheng (poor young male roles), and enjoyed the reputation of "the living image of Zhou Yu (a young general in the Three Kingdoms period) and Lü Bu (a young general in the Three Kingdoms period)".

Performance style: Xiao Sheng (young male) roles of the Ye School were famous for their singing technique, which was clear, sonorous, powerful, and combined vigour with suppleness. After absorbing the best techniques from talented actors, Ye Shenglan invented inverse Xipi (one of the modes used in Peking Opera music) tunes, emphasised the masculine charm of the Xiao Sheng roles, and created characters with rich and varied singing voices. Taking his performance of the character of Lü Bu as an example, Ye used Xipi tunes to present Lü Bu's arrogant and domineering characteristics in "Shooting The Halberd" (Yuan Men She Ji), whilst in "The White Gatehouse" (Bai Menlou), Ye used melancholy and plaintive tunes to suggest his sadness. In dance-acting technique, Ye Shenglan was particular about the details. When acting military generals, Ye looked strong and impressively heroic; when acting Wen Sheng (civilian young male) roles, he appeared handsome and graceful, and bookish. There were distinct differences in Ye's portrayal of Zhou Yu and Lü Bu, even though both belong to Zhiwei Sheng roles. The acting of Zhou Yu should emphasize his scholarly nature, whilst the presentation of Lü Bu should focus on showing his arbitrariness. The performance of Zhou Yu included having the chin slightly tucked in, eyebrows relaxed, eyes raised, lips slightly drawn, mouth closed, steps taken with control, each leg lifted with strength and lowered with control. In comparison, the performance of Lü Bu consisted of holding the head high, eyes widened, lips curled, mouth open, body relaxed, steps taken with swinging arms, legs lifted high but allowing them to float back down. Ye Shenglan also offered a summary for the use of shuixiu ('water sleeves'; long sleeves on costumes that are important to expression) with four key words: "heavy, light, supple, falling" (chen, yang, rou, luo). For example, Zhou Yu needs to be weighty, Lü Bu needs to be lighter, Liang Shanbo needs to be soft and supple, and Xu Xian needs express a sense of falling or lagging behind. Using programmatic techniques to represent the traits of a character is one of the main features of Ye's art. The masterpieces of Ye Shanglan included: "The Peach Blossom Fan" (Taohua Shan), "The Jade Hairpin" (Yu Zan Ji), "Phoenix Returning to its Nest" (Feng Huan Chao), "The White Snake" (Baishe Zhuan), "A Meeting of the Heroes" (Qun Ying Hui), "Dream in the Deserted Garden" (You Yuan Jing Meng), and "Zhou Ren Devoting His Sister-in-Low" (Zhou Ren Xian Sao).

2.3.

娃娃生

娃娃生扮演戏中儿童角色。娃娃生头戴孩儿发，身穿小茶衣、安安衣，念、唱用真嗓，唱腔不能用老生腔，而是混合了生、旦的唱腔，表演要带天真、可爱、伶俐的儿童稚气，如《铡美案》中的冬哥、春妹，《二堂舍子》中的秋儿，《三娘教子》中的薛倚哥，《锁麟囊》中的卢天麟等。娃娃生在剧团里很难找，常用身材娇小、扮相清秀的女演员来充任。

Wawa Sheng

Wawa Sheng (child roles) play children. Actors of the Wawa Sheng style their hair like children, wear a small Chayi (Tea Jacket) or An'anyi (Children's Jacket), and use a true voice when singing and speaking. The singing techniques of the Wawa Sheng combine those of the Xiao Sheng (young male roles) and the Dan (female roles); Lao Sheng techniques cannot be used for the singing of Wawa Sheng roles. The performance of Wawa Sheng should demonstrate the innocent, loveable and clever traits of children. Dong Ge and Chun Mei in "Executing Chen Shimei" (Zha Mei An), Qiu' er in "Saving Chen Xiang" (Er Tang She Zi), Xue Yige in "Sanniang Gives Her Son A Lecture" (Sanniang Jiao Zi), and Lu Tianlin in "The Jewellry Purse" (Suo Lin Nang) all belong to the Wawa Sheng role. It is often difficult to find children to play these roles, thus Wawa Sheng roles are often performed by petite and delicate younger actresses.

娃娃生——《锁麟囊》中的卢天麟

武生

3.

武生为京剧生行的一种，扮演武艺高强的男性角色。武生大多不挂髯口，以俊扮为主，少数勾脸的人物，称勾脸武生。武生唱、念用真嗓，念韵白，念白讲究吐字清晰，铿锵有力；表演上要求步子稳、把子准、档子狠。京剧形成初期，没有专设武生行当，分属于老生和小生行中。清光绪年间（1875 ~ 1908），俞菊笙、杨隆寿、黄月山、李春来等优秀武生演员的出现，使武生成为专行。早期武生表演以武技为主，后于 20 世纪初，武生演员杨小楼开创“武戏文唱”的表演风格，于开打表演中贯穿角色的思想感情和性格特点，使技与艺达到完美的统一，丰富了京剧武生行的表演艺术。

武生主要分长靠武生和短打武生两类。武生的流派有杨（小楼）派、盖（叫天）派等。

Wu Sheng

Wu Sheng (martial male roles) are one of main types of Sheng (male) roles in Peking Opera. They mostly take the role of young, physically strong males who are highly skilled in martial arts. Wu Sheng do not usually wear rankou (artificial beards), and mostly use "Junban" (handsome makeup, or, simple and light makeup) to decorate their faces (though there are a few minor roles that use more complex face painting, and these are called Goulian Wu Sheng, "martial male roles with a painted face"). When singing and speaking, Wu Sheng roles use a true voice and speak in yunbai (a stylized form of speech which brings a sense of decorousness to the character). When reciting lines, clear strong diction is required. When acting, it is important for actors to move steadily, to use stage weapons with accuracy, and fight energetically. In the early history of Peking Opera, there was no specific Wu Sheng role; characters were categorized within the Lao Sheng and

Xiao Sheng roles. During the reign of Guang Xu, at the end of Qing Dynasty, many outstanding Wu Sheng actors (such as Yu Jusheng, Yang Longshou, Huang Yueshan, and Li Chunlai) appeared on the stage, and as a consequence, the Wu Sheng became an identified role in Peking Opera performance. Early performances of the Wu Sheng focused mainly on the display of acrobatic fighting skills, but in early 20th century, famous Wu Sheng actor Yang Xiaolou created a new performing style: "martial drama with civilian singing" (wuxi wenchang) to enable the expression of the feelings of the characters alongside acrobatics. The total integration of the singing arts with fighting skills enriched the performance artistry of Wu Sheng roles.

Wu Sheng roles can be divided into two types: Changkao Wu Sheng (martial male roles who wear armour) and Duanda Wu Sheng (martial male roles who fight without weapons). The important schools of Wu Sheng are the Yang (Xiaolou) School, and the Gai (Jiaotian) School.

3.1.

长靠武生

长靠武生多扮演将帅角色，戴头盔，扎大靠，穿厚底靴，持长柄武器。这类武生，不但要求武功好，还要求突出将帅的风度和气魄，功架要优美、稳重、端庄。有些戏还要求表演细腻，并有一定唱、念的功夫。长靠武生演员在舞台上演戏有三种境界：“准、美、韵味儿。”这三种境界一级比一级难，而最后一级更是至关重要。长靠武生戏很多，如《长坂坡》《借东风》《回荆州》等中的赵云，《挑滑车》中的高宠，《战冀州》《反西凉》《战渭南》《两将军》《赚历城》等中的马超，《甘宁百骑劫魏营》中的甘宁，《青石山》中的关平等。

Changkao Wu Sheng

Changkao Wu Sheng (martial male roles who wear armour) mainly play young generals or chief commanders in the army who wear a helmet, hard armour, platform boots with thick soles, and bear long-handled weapons. This type of Wu Sheng is not only required to have a high degree of proficiency in martial arts, but also be able to express the imposing demeanour of generals. Therefore, the postures of Changkao Wu Sheng should be graceful, steady, and dignified. For some plays, an exquisite performance as well as strong singing and speaking skills are required. For actors of Changkao Wu Sheng roles, there are three fundamental elements required to perform well: "accuracy, beauty, and charm" (zhun, mei, yunwei'er). The second element is harder to achieve than the first, with the final element being the most difficult but also the most crucial. There are many Changkao Wu Sheng plays, such as Zhao Yun in "Changban Slope" (Changban Po), "Borrowing the East Wind" (Jie Dongfeng), "Returning to Jingzhou City" (Hui Jingzhou), Gao Chong in "Falling Chariots" (Tiao Hua Che), Ma Chao in "The Battle at Jizhou City" (Zhan Jizhou), "Rebelling in Xiliang" (Fan xiliang),

长靠武生——《两将军》中的马超

"The Battle at the South Bank of Weishui Rever" (Zhan Weinan), "Two Generals" (Liang Jiangjun), and "Capturing Licheng City by Deception" (Zhuan Licheng); Gan Ning in "Gan Ning Raiding Cao's Camp with A Hundred Cavalrymen" (Gan Ning Bai Ji Jie Wei Ying), and Guan Ping in "Blue Mountain" (Qingshi Shan).

3.2.

短打武生

短打武生多扮演草莽绿林英雄，着紧身短衣裤，穿薄底靴，用短兵器，大都表现步战，偏重近身特技。表演上重矫捷、灵活，以动作的干净利落、跌打翻滚、勇猛炽烈见长。如京剧《大破铜网阵》中的白玉堂、《十字坡》中的武松、《三岔口》中的任堂惠、《白水滩》中的十一郎、《闹天宫》中的孙悟空都属短打武生。

按照所穿服装来看，短打武生可以分为戴硬罗帽（穿抱衣抱裤或箭衣），戴软罗帽（穿抱衣抱裤、快衣快裤或箭衣），戴扎巾（穿箭衣）三类。如《恶虎村》《骆马湖》《连环套》等中的黄天霸戴硬罗帽。又如《打店》《打虎》《狮子楼》等中的武松，《三岔口》中的任堂惠，《四杰村》中的余千，《夜奔》中的林冲，《武文华》中的万君兆等，都是戴软罗帽的短打武生。还有戴扎巾、穿箭衣的，如《一箭仇》中的史文恭、《独木关》中的薛礼等，这类角色介乎长靠与短打之间，不仅重武功、功架，还重表演和唱、念。

Duanda Wu Sheng

Duanda Wu Sheng (martial male roles who fight hand to hand) mainly play young warriors, bandits, or outlawed heroes who wear tight-fitting jackets and trousers, ankle boots with thin soles, bear a short-

短打武生——《三岔口》中的任堂惠

handled weapon (e.g. a sword rather than a halberd), usually fight on foot (rather than horseback), and place a strong emphasis on hand-to-hand combat. In acting, agile and neat movements, and vigorous acrobatic tumbling is required. Examples of this role include Bai Yutang in "The Tactic of Breaking the Brass Net Battle Formation" (Da Po Tongwang Zhen), Wu Song in "Shizi Slope" (Shizi Po), Ren Tanghui in "At the Crossroads" (Sancha Kou), Shi Yilang in "The Baishui Shoal" (Baishui Tan), and Sun Wukong in "Havoc in Heaven" (Nao Tiangong).

Duanda Wu Sheng may be differentiated further into three types according to the costumes worn: roles wearing a yingluomao (hard silk hat) with tight-fitting jacket and trousers, or an archer's uniform; roles wearing a ruanluomao (soft silk hat) with tight-fitting jacket and trousers, or a combat outfit or archer's uniform; and roles wearing zhajin (bound kerchief) with an archer's uniform. For example, Huang Tianba in "Tiger Village" (E Hu Cun), "Luoma Lake" (Luoma Hu), and "The Strategem of Interlocking Rings" (Lian Huan Tao) wear yingluomao (hard silk hat), whereas Wu Song in "Fighting at the Inn" (Da Dian), "Beating a Tiger" (Da Hu), "Shizi Building" (Shizi Lou), Ren Tanghui in "At the Crossroads" (Sancha Kou), Yu Qian in "Sijie Village" (Sijie Cun), Lin Chong in "Fleeing at Night" (Ye Ben), Wan Junzhao in "Wu Wenhua" (Wu Wenhua) are ruanluomao (soft silk hat) roles. Example of roles wearing a zhajin (bound kerchief) include Shi Wengong in "An Arrow of Hatred" (Yi Jian Chou), and Xue Li in "Dumu Pass" (Dumu Guan). Some roles blend the performance techniques of the Duanda Wu Sheng (martial male roles who fight hand to hand) with those of Changkao Wu Sheng (martial male roles who wear armour), placing emphasis not only on a high level of dexterity in martial arts, fighting and dance-acting, but also on dance-acting, singing and recitation.

◎ 著名演员（一）

杨小楼（1878 ~ 1938）：京剧武生演员，杨派艺术的创始人。杨小楼在艺术上继承家学，自幼随父杨月楼练功学戏，后又师法俞菊笙、杨隆寿。杨小楼扮相俊逸，身段威武稳健，嗓音清脆洪亮，咬字清楚真切，行腔朴直无华。与梅兰芳、余叔岩并为 20 世纪二三十年代京剧三大代表人物，时称“三大贤”，享有“武生泰斗”“国剧宗师”的盛誉。

表演风格：杨小楼继承家学，风格融合俞菊笙和杨月楼两个流派之长，并吸收谭鑫培演技特点，博采众长，创立了“武戏文唱”的武生表演风格，逐渐形成独树一帜的“杨派”。杨派唱腔不使巧腔，铿锵和鸣，极富特色和韵味。其念白讲究尖团分明，感情真挚，体现剧情与人物的特定性格。在做功上，杨派武生注重武术功底，上下场、亮相优美大方。武打动作及武器运用突出一个“快”字，动作讲究洗练稳健，以简胜繁。武打步法准确灵敏，无空招废式。杨小楼善于用动与静、快与慢、含与露等对比手法加强表演效果。他所塑造的各类人物形象丰满且各具特色，相似扮相的人物表演绝不雷同。如马超与赵云，前者突出勇猛，后者则着重表现凝重。力求更恰当贴切地表现人物的性格，着力体现意境，追求神似，即“武戏文唱”的精髓所在。其代表作有《长坂坡》《战冀州》《甘宁百骑劫魏营》《挑滑车》《平阳关》《战宛城》等。

Famous Artist（Ⅰ）

Yang Xiaolou (1878 ~ 1938): was a famous Peking Opera actor of the Wu Sheng (martial male roles), and the founder of "Yang" School of WuSheng. Yang Xiaolou inherited his artistry from his family by learning from his father, Yang Yuelou. Later, he was apprenticed to Yu Jusheng and Yang Longshou. Yang's appearance on stage was handsome and athletic. His voice was resonant and clear with distinct enunciation. In singing, Yang used simple and unadorned melodies. Together with Mei Lanfang and Yu Shuyan, Yang Xiaolou was acknowledged as one of the three representative figures of Beijing Opera in 1920s and 1930s, known as "The Three Great Giants". He enjoyed a reputation as the "leading authority of Wu Sheng (martial male roles)" and "The Master of Chinese Opera".

Performance style: Based on the training he had undertaken with his family, Yang Xiaolou integrated the best techniques of Yu Jusheng and Yang Yuelou, and absorbed the characteristics of Tan Xinpei's acting to create his own unique approach to performing the Wu Sheng. This was known as "martial drama with civilian singing" (wu xi wen chang), and this became an independent school of performance known as the "Yang" School. The singing of the Yang School is relatively straightforward, but powerful, resonant and rich. In recitation, Yang took particular care over accurately distinguishing between jian (sharp) and tuan (round) tones in pronunciation (the two groups of tones come from two entirely different initial consonants in ancient Chinese), precisely reflecting the plot and the traits of each character. Wu Sheng from the Yang School focus on solid basic martial art skills, graceful and natural movements, and striking a strong pose (liang xiang) on the stage. The speed of acrobatic fighting action and the manipulation of weapons is highlighted in Yang's Wu Sheng performance. The movements need to be firm and steady, crisp rather than convoluted. When fighting, the feet must move accurately and nimbly. Yang Xiaolou placed emphasis on finding a contrast between moments of action and stillness, quickness and slowness, and implicitness and directness, to enhance the effect of his performance. The characters he created were full and unique; even those that were similar in appearance were not identical. Take Ma Chao and Zhao Yun (both were young and handsome male generals in the Three Kingdoms Period) for example. Yang's acting of the former centred on presenting his valour, while his acting of the latter focused on showing his dignity. The essence of "martial drama with civilian singing" lies in presenting the personality of the character, embodying the artistry that has been conceived, and achieving verisimilitude. Yang Xiaolou's masterpieces included: "Changban Slope" (Changban Po), "Falling Chariots" (Tiao Hua Che); "The Battle at Jizhou City" (Zhan Jizhou), "Gan Ning Raiding Cao's Camp with A Hundred Cavalrymen" (Gan Ning Bai Ji Jie Wei Ying), "Pingyang Pass" (Pingyang Guan), and "The Battle at Wancheng City" (Zhan Wancheng).

◎ 著名演员（二）

盖叫天（1888 ~ 1971）：京剧武生演员。盖叫天继承南派生创始人李春来的表演艺术，又广泛吸取其他表演艺术的长处并结合个人的条件加以发展，形成南派短打武生又一个重要流派，世称“盖派”。盖叫天擅长短打武生，曾经获得“第一勇猛武生”的称誉。

表演风格：盖派武生讲究人物的造型美，主张以外形动作表现人物的神情气质，将“立如松，坐如钟，卧如弓，行如风”的造型规则融入人物内在的精、气、神中。盖叫天的功架于稳练从容之中兼有脆率利落，其表演方法根据剧情及人物性格而不同，尤善以丰富变化的武打和造型表现不同的人物。同一人物在不同剧目中的塑造方式也不相同，如武松在《打虎》《狮子楼》《十字坡》《快活林》《蜈蚣岭》各剧目中的神态、武技都有所区别，比较清晰地勾勒了武松思想变化的脉络，树立了可信的英雄形象，使得盖叫天赢得“活武松”之誉。其代表作以“武松”（包括《打虎》《狮子楼》《十字坡》《快活林》等）及《三岔口》《一箭仇》等著名。

Famous Artist（Ⅱ）

Gai Jiaotian (1888 ~ 1971): was a famous actor of Wu Sheng (martial male roles). He was the founder of the "Gai" School of Wu Sheng. Gai Jiaotian inherited his art from Li Chunlai, founder of the southern style of performing art of Sheng (male roles). He then absorbed the best qualities of other styles of performance art which, when combined with his own approach, developed and formed an important school of southern style of Duanda Wu Sheng (martial male roles who fight hand to hand) known as "Gai" School.

Gai Jiaotian specialized in Duanda Wu Sheng, and enjoyed a reputation as "the first valorous Wu Sheng (martial male role)".

Performance style: Wu Sheng (martial male roles) of the Gai School emphasize the artistic modelling of characters, and stress the importance of displaying the internal temperament of characters through external movements. Performances expressed the mantra of "standing like a pine, sitting like a bell, lying like a bow, walking like the wind" into the inner energy, manner and spirit of the characters. Gai's acting was crisp and smooth, and differed according to the plot and personality of the character. Gai Jiaotian was expert at portraying different characters through different acrobatic fighting movements and stylistics. For the same role, different performance styles were used in different plays. Take Gai's acting of the famous hero Wu Song, for example. The demeanour of the character and the use of martial arts skills were distinguishable in different plays such as "Beating a Tiger" (Da Hu), "Shizi Building" (Shizi Lou), "Shizi Slope" (Shizi Po), "Regaining Happy Forests Pub" (Kuaihuo Lin), and "Wu Gong Mountain" (Wugong Ling). In doing so, audiences could discern how the character of Wu Song changed over time, making him seem more credible as a hero. Through this, Gai Jiaotian won the reputation of being "the living image of Wu Song". The classic plays of Wu Song starring Gai Jiaotian include: "Beating a Tiger" (Da Hu), "Shizi Building" (Shizi Lou), "Shizi Slope" (Shizi Po), "Regaining Happy Forests Pub" (Kuaihuo Lin), and "Wugong Mountain" (Wugong Ling). Gai Jiaotian's other representative masterpieces include: "At the Crossroads" (Sancha Kou) and "An Arrow of Hatred" (Yi Jian Chou).

第二章　旦行

旦行

旦行，京剧四大行当之一。旦扮演女性人物，分为正旦、花旦、花衫、刀马旦、武旦、老旦和彩旦等。

Dan

The Dan role is one of the four major role-type categories in Peking Opera. Dan roles are female characters, and according to the different ages and personalities of the characters, can be further divided into Zheng Dan, Hun Dan, Huashan, Daoma Dan, Wu Dan, Lao Dan, and Cai Dan.

正旦（青衣、青衫）

1.

按照传统，正旦在旦行里占据最主要的位置。因为扮演端庄、正派、严肃的中青年女性角色，且这类角色中相当一部分出身贫苦，常穿青色的服装，因而又名青衣、青衫。青衣的表演特点是以唱功为主，动作幅度较小，行动稳重，念韵白，唱功繁重，如《白蛇传·祭塔》中的白娘子、《祭江》中的孙尚香，以及《二进宫》中的李艳妃、《贺后骂殿》中的贺后等。也有些青衣戏是唱、念、做并重的，如《宇宙锋》中的赵艳容、《三击掌》中的王宝钏、《汾河湾》中的柳迎春、《宝莲灯》中的王桂英等。从服装上看，青衣扮演的人物主要穿青褶子，此外，还有女蟒、宫衣和帔。

Zheng Dan (Qingyi or Qingshan)

Traditionally, the Zheng Dan is the most important role in the Dan category. The characters of Zheng Dan are usually women who come from or find themselves in a lower status in society. As they often wear a costume that is blue-green (qing) in colour, the role is also sometimes called Qingyi (blue-green coat) or Qingshan (blue-green blouse). The Qingyi role places greatest emphasis on singing, with recitation in yunbai (a stylized form of speech that gives the character a sense of decorousness) and movement still significant, but to a lesser extent. The character of the White Snake in "The White Snake" (Baishe Zhuan), Sun Shangxiang in "Sacrifice to the River" (Ji Jiang), Li Yanfei in "Second Entrance to the Palace" (Er Jin Gong) and the Dowager Empress He in "The Dowager Empress He Curses the Throne" (Hehou Ma Dian) all belong to this role category. However, there are some roles where singing, dance-acting and recitation are equally important, such as Zhao Yanrong in "The Cosmic Blade" (Yuzhou Feng), Wang Baochuan in "The Third Palm Strike " (San Ji Zhang), Liu Yingchun in "Fen River Bend"

(Fenhe Wan) and Wang Guiying in "Lotus Lantern" (Bao Lian Deng). The Qingyi roles normally wear pleated skirts, as well as female mang robes, palace garments, and short embroidered capes.

正旦（青衣）——《武家坡》中的王宝钏

◎ 著名演员

程砚秋（1904 ~ 1958）：四大名旦之一，程派艺术的创始人。满族索绰罗氏，后改汉姓程。他幼时家贫，拜师学戏，历经艰辛。后来罗瘿公将他赎出。15 岁时程拜梅兰芳为师，后来，又向王瑶卿问艺，崭露头角。

他“倒仓”（发育期嗓音变化）后嗓音变化，一度被认为已无前途。可是程砚秋并不失望悲观，在王瑶卿等师友的指点和帮助下，结合自己的嗓音特点，另辟蹊径，扬长补短，创造出一种幽咽婉转的新腔。

程擅演悲剧，代表剧目有《锁麟囊》《荒山泪》《青霜剑》《文姬归汉》等。《锁麟囊》集程派唱腔艺术之大成，最令人倾倒。

Famous Artist

Cheng Yanqiu (1904 ~ 1958): was one of the "Four Great Dan Actors", though the surname Cheng was adopted later in his life. He came from a poor family, and was sold to a troupe as an apprentice to learn drama, suffering many hardships during his training. Later, Luo Yinggong helped to release him from the bondage of his apprenticeship. By the age of fifteen, he was a student of Mei Lanfang. Later, he also asked to learn from Wang Yaoqing, and then he began to stand out as a performer.

Once Cheng's voice had broken, it became dull and rough, and it was felt that he had no future in the theatre. However, Cheng Yanqiu was not disappointed or pessimistic. With guidance from Wang Yaoqing, other teachers, and friends, he developed his own characteristic sound and vocal technique.

Cheng was good at tragedy, as is demonstrated by his appearance in "The Jewelley Purse" (Suo Lin Nang), "Shedding Tears on the Barren Mountain" (Huangshan Lei), "Qingshuang Sword" (Qingshuang Jian), and "Wen Ji Returns to the Han" (Wenji Gui Han). Among them, The Jewelley Purse represents the most outstanding achievement of the Cheng School's style of singing.

His representative works included: "The Jewelley Purse" (Suo Lin Nang), "Shedding

Tears on the Barren Mountain" (Huangshan Lei), "Qingshuang Sword" (Qingshuang Jian), "Wen Ji Returns to the Han" (Wen Ji Gui Han).

花旦

2.

花旦，是旦行中的一支，主要扮演天真活泼或身份较低的年轻女子，多穿短衣裙，重视念白和做，如《西厢记》中的红娘。

花旦又分闺门旦、玩笑旦、泼辣旦、刺杀旦。闺门旦，扮演的是还没有出嫁的少女，性格内向、腼腆，与正旦接近。玩笑旦，扮演的是喜剧、闹剧中爱说爱笑、好打好闹的人物，如《小上坟》《打花鼓》中的女性。泼辣旦，扮演的是举止放荡、说话锋利的女性，如《翠屏山》中的潘巧云、《巴骆和》中的马金定。关于刺杀旦，有两种认识：一是说刺杀别人的；二是说被别人刺杀的。前者如《双钉记》中的白金莲，后者如《战宛城》的邹氏等。

Hua Dan

Hua Dan roles portray naïve and sprightly young girls with a lower social status. They wear relatively shorter skirts, and emphasise the skills of dance-acting and recitation in performance. The role of Hong Niang in "The Romance of the Western Chamber" (Xixiang Ji) offers a vivid example of this role.

The Hua Dan can be further subdivided into: Guimen Dan, Wanxiao Dan, Pola Dan and Cisha Dan. Guimen Dan is for unmarried young women who have shy and introverted personalities, and is similar to the Zheng Dan role. The Wanxiao Dan portrays comic, chatty, and farcical female roles, such as the female characters in "Visiting the Little Grave" (Xiao Shang Fen) and "Beating the Flower Drum" (Da Huagu). Pola Dan, literally means "Shrewish Dan" and, as suggested by its name, is for young girls who are somewhat profligate and rather sharp talking. Pan Qiaoyun in "Cuiping Mountain" (Cuiping Shan) and Ma Jinding in "Reconciliation between Ba and Luo" (Ba Luo He) are performed by the Pola Dan. There is one more division called the Cisha Dan. "Cisha"

花旦——《西厢记》中的红娘

literally means "to assassinate" and the role can be interpreted in two ways: women who seek to assassinate others, and women who were themselves assassinated. For instance, Bai Jinlian in "The Story of the Pair of Nails" (Shuangding Ji) is a Dan character who seeks to kill others, whilst Zoushi in "Wancheng Battle" (Zhan Wancheng) is a woman who is killed.

◎ 著名演员

荀慧生（1900 ~ 1968）：自幼家贫，7 岁时被卖给河北梆子艺人为徒，学花旦，后到北京学京剧，与尚小云同科。他相继师从王瑶卿、吴菱仙，进步很快。1925 年 10 月 2 日，荀与余叔岩合演《打渔杀家》，从此用荀慧生之名。

荀兼擅青衣、花旦、刀马旦，善于塑造天真可爱、活泼热情的少女形象，表演亲切自然，具有生活化的特点。他吸收梆子的精华，丰富了京剧艺术。他在念白方面首创“谐白”（结合京白和韵白）的特殊念法，做功尤为出色，增加了许多细腻妩媚且生活化的小动作，配合眼神的运用，表现少女和少妇的心理活动。他还创造了一系列独特的身段动作。

代表作品为《拾玉镯》。

Famous Artist

Xun Huisheng (1900 ~ 1968): was from a poor family, and was sold at the age of seven to a Hebei Bangzi Opera troupe to learn the Hua Dan role. Later he went to Beijing to learn Peking Opera and studied alongside Shang Xiaoyun. He studied under Wang Yaoqing and Wu Lingxian successively, making great progress. On 2 October 1925, Xun was on stage with Yu Shuyan in "The Fisherman's Revenge" (Da Yu Sha Jia), and thereafter went by the name of Xun Huisheng.

Xun was skilled in the Qingyi, Hua Dan and Daoma Dan roles, but was especially good at depicting innocent, lively, and warm-hearted young girls. His performances on stage were affectionate, natural and realistic. He enriched Peking Opera by absorbing the essence of Bangzi Opera into his work. In the art of recitation, he created a new style, called "xiebai" (literally "harmonic speaking"), which was a special style that integrated the Beijing dialect used on stage (jingbai) with stylized speech (yunbai). His dance-acting was especially remarkable. He added many delicate, charming and life-like movements. Through the use of his eyes, he was able to express the psychology of young girls and young women. He also created a series of new choreographic movement patterns.

His representative works included: "Picking up the Jade Bracelet" (Shi Yuzhuo).

花衫

3.

花衫是京剧旦行里的一个重要的行当。在花衫形成之前，一个演员一般不能兼演青衣和花旦两个行当。为了丰富旦角的表现力，契合日益多元的社会生活和复杂多样的人物性格，从而提升戏曲表演的观赏性，王瑶卿融合了青衣的沉静端庄、花旦的活泼灵巧、武旦的武打功架，创造出一种唱、念、做、打并重的新行当，花旦、青衫的结合体——花衫。它要求演员多知多能，全面把握人物角色，准确刻画人物性格。梅兰芳是最具代表性的演员。

Huashan

Huashan is of great importance to the Dan category. Before the creation of this role, a performer was not allowed to act both the Hua Dan and Qingyi roles. In order to enrich the Dan role, and to enable Peking Opera to portray the diversity of society through more complex characters, Wang Yaoqing created this role category. He integrated the calm and dignified air of the Qingyi role, with the dexterity of the Hua Dan role, and the martial arts movements of the Wu Dan role. For Huashan actors, singing, recitation, dance-acting and fighting skills are all equally important. Actors of this role need to be good all-rounders, and be able to understand and express characters. Mei Lanfang was the greatest Huashan actor.

花衫——《贵妃醉酒》中的杨贵妃

◎ 著名演员

梅兰芳（1894 ~ 1961）：“四大名旦”之首，是卓越的京剧艺术大师，是举世公认的中国戏曲艺术的杰出代表。他本名澜，出身梨园世家。梅兰芳 4 岁丧父，8 岁学戏，11 岁登台演出。

梅兰芳的艺术生涯长达50余年。梅兰芳对京剧的编、导、表演、唱腔、伴奏、化妆、服装、灯光、道具等方面都进行了一系列的改革。比如在服装扮相方面，梅参考古代仕女画设计了裙服和发型。

代表作品有《宇宙锋》《贵妃醉酒》《霸王别姬》《天女散花》《太真外传》《穆桂英挂帅》等。

Famous Artist

Mei Lanfang (1894 ~ 1961), the most important of the "Four Great Dan Actors", is an outstanding artistic master of Peking Opera, and also universally acknowledged as the world-wide representative of Chinese theatre arts. He came from an acting family, his original name being Lan. When Mei Lanfang was four, his father died, and he started learning Peking Opera when he was eight years old. He made his debut on stage aged eleven.

Mei Lanfang's artistic career lasted more than fifty years. He carried out a series of reforms in the areas of Peking Opera composition, directing, performing, singing, music, makeup, costume, lighting, and props. For example, for costumes, Mei consulted historical artworks to create clothing and hairstyles in the appropriate historical style.

Mei Lanfang's representative works included: "The Cosmic Blade " (Yuzhou Feng), "The Imperial Concubine Becomes Intoxicated" (Guifei Zui jiu), "The King Bids Farewell to His Favourite" (Bawang Bie Ji), "The Heavenly Maiden Scatters Flowers" (Tiannü San Hua), "The Tale of Taizhen" (Taizhen Waizhuan), and "Mu Guiying Takes Command" (Mu Guiying Gua Shuai).

4. 老旦

老旦扮演老年妇女人物，其扮相、身段与台步都要突出老年人的特点。唱、念都用本嗓，但唱腔具有女性婉转迂回的韵味，突出女性特征，这种声腔在京剧的行话里叫“雌音”。从表演技艺来说，老旦一般可以再划分为唱工老旦和做工老旦两类，如《钓金龟》中的康氏、《赤桑镇》中的吴妙贞、《望儿楼》中的窦太真、《遇皇后・打龙袍》中的李后等，都属于唱工老旦；《清风亭》中的贺氏、《西厢记》中的崔老夫人、《得意缘》中的狄母等，都属于做工老旦。在 20 世纪 80 年代自地方戏移植的《对花枪》中，老旦有了披靠、开打的技艺，丰富了老旦行当的表现能力。

Lao Dan

Lao Dan roles portray old women characters. Their appearance, posture and gait are characteristic of elderly women. As its hallmark, the actors sing and recite in their natural voices, which has a pleasant, gentle, melodious and sweet sound, which evokes femininity. Such vocal music is called the "feminine voice" (ci yin) in the jargon of Peking Opera. Lao Dan roles may be further divided into two categories based upon a different emphasis on skills in performance: the Changgong Lao Dan (Singing Lao Dan) and Zuogong Lao Dan (Acting Lao Dan). Mrs. Kang in "The Tale of a Tortoise" (Diao Jingui), Wu Miaozhen in "Chisang Town" (Chisang Zhen), Dou Taizhen in "A Mother's Longing for Her Son to Return Home" (Wang Er Lou), and Empress Li in "Encounter with the Empress" (Yu Huanghou) all belong to the Singing Lao Dan role. Mrs He in "Breeze Parilion"(Qingfeng Ting), The Dowager Cui in "The Romance of West Chamber" (Xixiang Ji), and Mother Di in "Darts for Males and Females" (Deyi Yuan) are performed by the Acting Lao Dan. In a regional drama from the 1980s called "Silver Spear" (Dui Huaqiang), a Lao Dan character wore armour military costume and began to use fighting skills on stage, which enriched the performance of this role.

◎ 著名演员（一）

龚云甫（1862～1932）：近代京剧老旦演员。初习老生，后改演老旦。进入鸿庆班，在广和楼演唱，声誉极佳。常与梅兰芳、杨小楼等名家合作，并三次应邀到上海以老旦挂头牌演出。在吸收借鉴的基础上，龚云甫将老旦的唱腔、念白和身段加以规范，老旦的表演被推向了一个新水平。自龚云甫起，老旦成为一个独立的表演行当。他的唱、念讲究音韵，嗓音清脆、沧桑，行腔悠扬婉转，抑扬顿挫，起落有序。在表演上注意切合剧中人物身份，尤其善于运用面部表情，风格独特。龚云甫的演唱进而发展为影响巨大的老旦流派，世称龚（云甫）派。《钓金龟》《徐母骂曹》《太君辞朝》《打龙袍》等是他经常上演的剧目。

Famous Artist（Ⅰ）

Gong Yunfu（1862 ~ 1932）was a more recent Lao Dan actor in Peking Opera. He started learning as Lao Sheng and then changed to train as a Lao Dan. After entering the Hongqing Troupe, he performed at Guanghe Pavilion and gained a good reputation. The actors who worked with him were all famous, such as Mei Lanfang and Yang Xiaolou. Gong was invited to perform in Shanghai three times as the leading actor. As a result of his careful absorption of training, he standardized singing, recitation and dance-acting, which elevated the performance of the Lao Dan to a new artistic level. It is because of Gong Yunfu that Lao Dan has become an independent role type in the art Peking Opera. He placed emphasis on the clarity of singing and recitation, giving a crispness to his voice that was capable of great expression, and was melodious and controlled. In acting, he ensured that his embodiment of character was closely matched to the identity of the role he was playing,

老旦——《徐母骂曹》中的徐母

and he was particularly good at using facial expressions. His style was unique. Because his singing became so influential, his style evolved into the Gong School.

His representative works included: "The Tale of a Tortoise" (Diao Jingui), "Mother Xu Scolds Cao Cao" (Xumu Ma Cao), "Bidding Farewell to the Court" (Taijun Ci Chao), and "Beating the Dragon Robe" (Da Longpao).

◎ 著名演员（二）

李多奎（1898 ~ 1974）：9 岁开始学习河北梆子和京剧老生，11 岁登台。变声后改习京剧老旦。1923 年首演《钓金龟》，一举成名。李多奎博采众长并结合自己的条件，成为唱工老旦代表人物，至 20 世纪 40 年代自成一家，创立了古朴大方、苍迈挺拔，韵味醇厚的老旦新腔，世称“奎派”， 又称“李派”。1961 年，他排演了全部由自己创腔的《赤桑镇》，成为京剧的保留剧目。1963 年，拍摄电影《铡美案》。一生大约录制 100 张唱片。

代表作品有《钓金龟》《望儿楼》《赤桑镇》《四郎探母》《 铡美案》《遇皇后》《龙凤呈祥》《辕门斩子》。

Famous Artist（Ⅱ）

Li Duokui（1898 ~ 1974）began studying Hebei Opera and the Lao Sheng role in Peking Opera at the age of nine, making his stage debut aged eleven. He switched to the Lao Dan in Peking Opera after his voice broke. He first performed "The Tale of a Tortoise" (Diao Jingui) in 1923, and instantly became famous. Li Duokui drew upon the skills he had learnt and combined them with his own artistry to become the most significant Lao Dan singer. In the 1940s he established what is commonly known as the Kui School, also known as the Li School, which is famous for its simple and elegant vocal music, and is full of charm. In 1961, he performed "Chisang Town" (Chisang Zhen) which he created entirely by himself. The play is now part of the standard Peking Opera repertoire. In 1963, he starred in the film "Executing Chen Shimei" (Zha Mei An). During his lifetime, he recorded almost 100 LPs.

His representative works included: "The Tale of a Tortoise" (Diao Jingui), "Mother's Longing for her Son to Return home" (Wang Er Lou), "Chisang Town" (Chisang Zhen), "Si Lang Visits His Mother" (Silang Tan Mu), "Executing Chen Shimei" (Zha Mei An), "Encounter with the Empress" (Yu Huanghou), "Grand Wedding in the Enemy Camp" (Long Feng Cheng Xiang), and "General Yang Liulang Executes His Own Son" (Yuanmen Zhan Zi).

武旦

5.

武旦扮演精通武艺、英勇善战的女性人物，如武将、江湖中的各类女侠等。穿短衣裳，一般不骑马（也有骑马的，比较少），重在武功与念白。有的还以武打和跌扑取胜，不太注重演唱等表演内容。例如《杨排风》中的杨排风、《泗州城》中的水母、《打店》中的孙二娘、《无底洞》中的白鼠精、《摇钱树》中的张四姐、《三岔口》中的店主婆等，都属于武旦。

武旦在表演上重武打和绝技的运用。武旦的特点有二：一是跷功，即脚上绑木质小脚，模拟古代妇女缠足行走，而且还要完成一些高难度动作，如走凳子、踢石子。它十分考验演员的功力。二是打出手，即以一个角色为中心，另有几个角色抛扔武器，相互配合，做抛、掷、踢、接武器的特技表演，形成种种丰富繁难的惊险场面。根据剧情需要和演员水平，可由三人、五人、七人、九人表演，所用枪也由两杆至十几杆不等。打出手有专用的锣鼓伴奏，以烘托气氛。

Wu Dan (Warrior Dan)

Wu Dan are female roles who are proficient in martial arts and strong in character, such as generals, or women who have acquired martial arts (jianghu). They generally wear shorter costumes and do not ride horses (there is some horse riding, but to a lesser extent). In performance, actors emphasize acrobatic fighting and recitation. Some are also good at martial arts and tumbling, but do not pay so much attention to singing skills. For instance, Yang Paifeng in "Yang Paifeng" (Yang Paifeng), the Water Goddess in "Sizhou City" (Sizhou Cheng), Sun Erniang in "Fighing at the Inn" (Da Dian), the Jade Rat in "The Abyss" (Wudi Dong), Zhang Sijie in "Money Tree" (Yao Qian

武旦——《锔大缸》中的王大娘

Shu), the Landlady in "At the Crossroads" (Sancha Kou) are all Wu Dan characters.

Wu Dan roles emphasize martial arts and special stunts. There are two special kinds of stunt. One of these is stilt-walking. To imitate the bound feet of women from past times, the actor will step on a pair of wooden stilts, which are tapered and small. During the performance, the actors can only walk on their toes. When walking, the actor not only imitates the gait of women with bound feet, but also accomplishes difficult movements, such as stepping onto a stool, or kicking a pebble. This really tests the skills of the actor. The other special stunt is called "Da Chu Shou" and involves a number of actors tossing, casting, kicking and receiving weapons across the stage with one leading actor positioned in the centre. It is a breath-taking dance-like performance. According to the needs of the plot and the skill of performers, three, five, seven or nine people may be involved, and the number of spears they use varies from two to over ten. Da Chu Shou acrobatic fighting is accompanied by special drums and gongs, which adds to the excitement.

6. 刀马旦

京剧中身穿铠甲、头带首盔、骑马鏖战的擅长武艺的女性人物，持长兵器，如刀，被称为刀马旦。实际上是武旦和花衫融合的一种行当。刀马旦一方面要有很好的武功，同时还长于做功，而且有时候说白、功架也很重要。如《穆柯寨》《穆天王》《破洪州》中的穆桂英，《七星庙》《佘塘关》中的佘赛花，《棋盘山》中的窦仙童，《马上缘》《三休樊梨花》中的樊梨花，《珍珠烈火旗》中的双阳公主，《刘金定》中的刘金定，《扈家庄》中的扈三娘等。

刀马旦——《珍珠烈火旗》中的双阳公主

Daoma Dan

Daoma Dan is a female character who wears armour, and is also good at horse-riding, sword-wielding, and manipulating weapons. That is why the role is called "Daoma Dan" – "dao" means "sword" and "ma" means "horse". The Daoma Dan is a blend of the Wu Dan and Huashan roles. Daoma Dan actors must have excellent martial arts skills, and also excel in dance-acting, though sometimes recitation is also important. Examples of this role include Mu Guiying in "Muke Village" (Mu Ke Zhai), "The Heavenly King" (Mu Tianwang), and "Lift the Siege of Hongzhou" (Po Hongzhou); She Saihua in "Temple of Seven Stars" (Qixing Miao) and "Shetang Pass" (Shetang Guan), Dou Xiantong in "Mount Qipan" (Qipan Shan), Fan Lihua in "Romance On Horseback" (Ma Shang Yuan) and "Fan Lihua Deserted" (San Xiu Fan Lihua), Princess Shuangyang in "The Flag of Pearl and Fire" (Zhenzhu Liehuo Qi), Liu Jinding in "Liu Jinding" (Liu Jinding) and Lady Hu in "Hu Village" (Hu Jia Zhuang).

彩旦

7.

彩旦俗称丑婆子，因为面部化妆和服装穿戴刻意突出花哨的色彩而得名，也称丑旦、玩笑旦，即女丑角。彩旦的唱、念用本嗓，说京白，重做功，表演、化妆都很夸张，主要扮演滑稽、诙谐的女性喜剧人物。在舞台上，彩旦嬉笑怒骂，逗乐搞笑，往往最能烘托气氛，使台下观众兴致浓厚。如《凤还巢》中的程雪雁、《西施》中的东施、《反徐

彩旦——《拾玉镯》中的刘媒婆

州》中的花婆、《四进士》中的万氏、《铁弓缘》中的陈母、《拾玉镯》中的刘媒婆等。

Cai Dan

Cai Dan (literally "Coloured Dan" role), also commonly known as Choupozi (literally "ugly old woman") is famous for its facial make-up and colourful clothing, which explains its name. Cai Dan is also called Chou Dan ("Clown Dan") or Wanxiao Dan ("Joking Dan"). Cai Dan actors both sing and recite in their natural voices, and the speech is mainly in jingbai, a form of speech similar to the everyday Beijing dialect. Cai Dan performances emphasize posture and movement. The acting and make-up are exaggerated, thus, the role specializes in acting funny, humorous comic characters. On stage, the Cai Dan laughs mischievously and curses angrily, embellishing the atmosphere and maintain the interest of the audience. Examples of Cai Dan include Cheng Xueyan in "The Phoenix Returns to its Nest" (Feng Huan Chao), Dong Shi in "Xi Shi" (Xi Shi), Huapo in "Rebellion in Xuzhou" (Fan Xuzhou), Mrs Wan in "The Lawsuit" (Si Jinshi), Mother Chen in "The Iron Bow" (Tiegong Yuan), and Matchmaker Liu in "Picking Up The Jade Bracelet" (Shi Yuzhuo).

第三章　净行

净行

净行是中国戏曲行当的一种，净主要扮演性格、品质或相貌不同于一般、有突出特征的男性人物。净面部重施油彩，勾画各式具有复杂图案的脸谱，俗称“花脸”“花面”。净的表演特色：声音洪亮宽阔，演唱风格粗豪浑厚，动作大开大阖，顿挫鲜明。饰演的人物个性或是粗犷豪迈，或是阴险凶残，或是刚烈耿直，或是鲁莽诚朴；上至帝王将相，下到庶民屠夫；既有正面人物，也有反面人物。

净分为正净、副净和武净三类。

Jing (painted face)

Jing (painted face) is one of the role types in Peking Opera. Jing actors usually play male roles that are distinctive in personality, quality or appearance. Jing roles, normally known as Hualian or Huamian (literally "flower face", but usually translated as "painted face"), are characterized by having brightly painted faces, often in complex patterns.

Jing roles have a sonorous voice, a vigorous singing style, and a large range of movements. The characters can be bold and heroic, sinister and cruel, honest or reckless, good or bad, and are drawn from all social classes.

Jing can be subdivided into three different types: Zheng Jing, Fu Jing and Wu Jing.

正净

1\. 又名“大花脸”“大花面”，简称“大面”，以唱功为主，故又称“唱工花脸”。《二进宫》剧中的徐彦昭，是典型的唱工花脸，因他手持铜锤，所以人们就把“铜锤”作为唱工花脸的代名词。此外，还有很多的包公戏，如人们常见的《打龙袍》《铡美案》等，大多唱功繁重，包公又是勾画黑色脸谱的代表人物，所以“黑头”也就成为唱工花脸的又一个代名词。其他的如《姚期》中的姚期、《将相和》中的廉颇，还有《牧虎关》中的高旺等，都是以唱功为主的花脸，都属于正净的范畴。

正净的流派有金（少山）派、裘（盛戎）派等。

Zheng Jing (positive painted-face male role)

Zheng Jing is known by many other names, including Da Hualian, Da Huamian (both meaning "large painted face"), and Da Mian ("large face") for short. It is also referred to as Changgong Hualian (literally "painted face with singing skills") because Zheng Jing roles focus on art of singing in performance. A typical Changgong Hualian role is Xu Yanzhao in "Entering the Palace for the Second Time" (Er Jin Gong). Because the character holds a copper hammer in his hand, "copper hammer painted face" (Tongchui Hualian) has also become a synonym for "the singing painted face" (Changgong Hualian). In addition, the character of Bao Gong, who appears in many plays, including "Beating the Emperor's Robe" (Da Long Pao) and "Executing Chen Shimei" (Zha Mei An), is the representative character for the style of painted face that is predominantly black in colour. Therefore, "black face"(hei tou) is another synonym for the Changgong Hualian role. Other examples of Zheng Jing roles include Yao Qi in "Yao Qi" (Yao Qi), Lian Po in "Reconciliation between the General and the Premier" (Jiang Xiang He), and Gao Wang in "Muhu Pass" (Muhu Guan).

There are two representative Schools of the Zheng Jing roles: the Jin (Shaoshan) School and the Qiu (Shengrong) School.

◎ 著名演员（一）

金少山（1890 ~ 1948）：铜锤花脸演员。金少山确立了京剧史上第一个唱、念、做、打全面发展的、完整的花脸流派，世称“金（少山）派”。风格雄浑豪放，大气磅礴。以唱功见长。

金少山是著名京剧演员金秀山之子。幼年随父学艺，在继承父亲铜锤唱腔的基础上，吸收了架子花脸的唱、做，突破了两者严格的分工界限。在吐字、行腔的技巧运用方面也较一般铜锤花脸细致，形成了自己的风格。

金少山嗓音洪亮，他的发声方法是科学的，尤其是头腔、胸腔共鸣结合得非常好。金少山改变了前人一些笨拙的唱法，改造了那种接近喊叫的声音。他的新唱法具有工稳、流畅的特点，代表着 20 世纪三四十年代花脸的最高演唱水平。

代表剧目有《锁五龙》《草桥关》《白良关》《飞虎山》等。

Famous Artist（Ⅰ）

Jin Shaoshan (1890 ~ 1948) was an actor of Tongchui Hualian ("singing painted face") roles. Jin was the first to establish the integration of the four skills of Peking Opera (singing, recitation, dance-acting and fighting) for the painted face roles, which became known as the Jin School. The School is famed for its vigorous and magnificent performance style, and for its accomplishment in the skills of singing.

When he was a child, Jin Shaoshan started learning performing skills from his father, Jin Xiushan, who was a well-known actor of Peking Opera. Jin learnt the Tongchui Hualian tunes from his father, and absorbed the arts of singing and dance-acting, which he then

正净（花脸）——《铡美案》中的包拯

synchronised, breaking down the strict divisions between the two skills. In terms of the interpretation of lines and articulation of words, Jin developed a more elaborate style than that of other Tongchui Hualian actors.

Jin's voice was loud and clear. His vocal techniques were scientific, especially in finding a means of using the head and chest cavities in combination to produce resonance. He improved pre-existing vocal methods, which seemed to be shouting. In the 1930s and 1940s, Jin's new approach to singing, which was more steady and fluent, elevated the singing techniques of the painted face roles to their highest level.

His representative works included: "Execution of Xiongxin" (Suo Wu Long), "Caoqiao Pass" (Caoqiao Guan), "Bailiang Pass" (Bailiang Guan), and "Feihu Mountain" (Feihu Shan).

◎ 著名演员（二）

裘盛戎（1915 ~ 1971）：铜锤花脸演员。裘盛戎吸收各派唱、做精华，形成新唱法，在“刚”中注入了“柔”的成分，在花脸声腔发展的历程中，为金少山之后的第二座里程碑，世称“裘派”。

裘盛戎的父亲裘桂仙也是著名的花脸演员。裘盛戎 8 岁开始随父亲学戏，他是铜锤和架子花脸两门抱，以铜锤为主。他嗓音既有高昂激越的一面，又有迂回委婉的一面，刚柔并济。他能结合自己的嗓音特点，从表现人物出发，在词义和字声方面下功夫，塑造了不同的艺术形象，有很强烈的艺术感染力。

裘盛戎在长期的舞台实践中能博采众长，经过不断钻研，创造了独树一帜的“裘派”唱腔。他创造的唱腔继承了净角传统唱腔雄浑豪放的风格，又融合了老生、青衣唱腔中低回婉转的抒情特色，形成了韵味醇厚、含蓄细腻的风格。裘盛戎为京剧事业发展做出了很大贡献。

代表剧目有《法门寺》《遇皇后》《牧虎关》《铡美案》等。

Famous Artist（Ⅱ）

Qiu Shengrong (1915 ~ 1971), actor of the Tongchui Hualian roles. Qiu absorbed the best elements of singing and dance-acting from each School to form the Qiu School. He established a new method of singing, which injected softness into the voice of the Jing roles, which is the second milestone in the development of vocal techniques after Jin Shaoshan.

Qiu Guixian, Qiu Shengrong's father, was also a famous actor of the painted face role. At the age of eight, Qiu Shengrong started learning performance from his father. He became a Tongchui Hualian and was also good at the Jiazi Hualian (posture painted face) role. His voice could express high-spiritedness and sense of agitation, or be softer and more euphemistic. Qiu carefully studied the meaning and tones of the recited parts and, using his vocal techniques to good effect, could express each character vividly, making them aesthetically appealing.

After a long career on the stage, Qiu was able to understand what pleased audience, and after continuous research, he was able to create a unique singing style for the Qiu School. Drawing upon the traditional melodies of the Jing roles, he combined his own melodies with the mild and lyrical ones of the Lao Sheng and Qingyi roles, creating a mellow and subtle style. His artistic achievements made a great contribution to the development of Peking Opera.

His representative works included: "Famen Temple" (Fa Men Si), "Meeting the Empress" (Yu Huanghou), "Muhu Pass" (Muhu Guan), and "Executing Chen Shimei" (Zha Mei An).

副净

2.

又称架子花脸。顾名思义是以功架（动作、造型）表演为主，当然也得有唱功基础。如《盗御马》中的窦尔敦，既要有精湛的武功，又要善于表演、念白，还要有唱功。一个优秀的架子花脸演员，应该是文武全才。架子花脸可表现的人物很多，如张飞、李逵，以及焦赞和曹操等。还有如《失街亭》中的马谡、《取洛阳》中的马武、《审李七》中的李七、《李慧娘》中的贾似道、《群英会》中的黄盖等，都属于架子花脸范畴。

特别值得一提的是曹操的脸谱，一般不用油彩，而用水粉加上一些墨笔勾成水白脸，俗称"粉脸"。历史上的很多奸臣形象，像赵高、董卓、秦桧等，都勾水白脸，但以曹操为典型。这样的扮演方式已经达成共识，所以京剧舞台上凡勾水白脸的，就一定是坏人。而勾水白脸的人物，都由架子花脸来应工扮演。

二花脸也属于副净范畴，是架子花脸的一种，不过戏比较少。虽然也勾同样的脸谱，可是表演风格有时近似丑角，有时候还扮演一些诙谐狡猾的角色。如《算粮》中的魏虎、《法门寺》中的刘彪等。

副净的流派有郝（寿臣）派、侯（喜瑞）派等。

Fu Jing (supporting painted-face male role)

Fu Jing is one of the subdivisions for Jing roles, and is also known as Jiazi Hualian ("posture painted face"). In addition to the art of singing, the performance of Fu Jing roles is focused on movement and postures. Dou Erdun in "Stealing the Imperial Horse" (Dao Yu Ma) is a good example of a Fu Jing role because the actor must be good at martial arts skills, acting, and recitation, as well as singing. This is to say that the best actors of Jiazi Hualian roles must be versatile. There are many famous characters of Jiazi Hualian, such as Zhang Fei, Li Kui, Jiao Zan and Cao Cao. Other examples include Ma Su in "Loss of Jieting" (Shi

副净（架子花脸）——《盗御马》中的窦尔敦

Jieting), Ma Wu in "The Capture of Luoyang City" (Qu Luoyang), Li Qi in "Interrogating Li Qi" (Shen Li Qi), Jia Sidao in "Li Huiniang" (Li Huiniang), and Huang Gai in "The Meeting of the Heroes" (Qun Ying Hui).

The character of Cao Cao is one Jing role worth highlighting, as his face is not painted with oil-based make-up like other Jing roles. Rather, this role uses water-based powder paint as a base, on top of which ink pen outlines are applied. This creates a "water white face" (shui bai lian), more commonly known as "powder face" (fen lian). There have been many treacherous officials across Chinese history, such as Zhao Gao, Dong Zhuo and Qin Hui, all of whom are depicted with "water white face" make-up. It has now become a convention for audience to understand that the character must be an evil one who has the "water white face" make-up. Characters who wear the "water white face" make-up are all played by the Fu Jing roles.

Er Hualian is another name often used for the Jiazi Hualian, but its performing style is in fact different, being in some ways similar to the Chou (clown) roles, as the characters are sometimes humorous. Examples of the Er Hualian roles include Wei Hu in "Demanding Provisions" (Suan Liang), and Liu Biao in "Famen Temple" (Famen Si).

There are two important schools of the Fujing roles: the Hao (Shouchen) School and the Hou (Xirui) School.

◎ 著名演员（一）

郝寿臣（1886~1961）：架子花脸演员。郝寿臣创立的郝派，与金（少山）派、侯（喜瑞）派鼎足而立，是 20 世纪 20 年代以来净行三大流派。

郝寿臣在两百多出戏中扮演过主要和次要角色。郝派艺术以“架子花脸铜锤唱”为特点，功架精美，表演浑然一体，并善于刻画人物的性格，对架子花脸表演艺术有很多革新创造。郝寿臣还设计了鲁智深的螳螂眉、和尚眼，粉红双颊脸谱和袒胸裸腹的扮相，突出其豪爽朴实的形象。

郝寿臣还先后创演了《荆轲传》《桃花村》《牛皋招亲》等新戏。

代表剧目有《打曹豹》《盗御马》《野猪林》《斩马谡》等。

Famous Artist（Ⅰ）

Hao Shouchen (1886 ~ 1961) was an actor of Jiazi Hualian ("posture painted face") roles. The Hao School, founded in 1920s, was one of the three important Schools of Jing roles, the other two being the Jin School and Hou School.

Hao Shouchen mainly acted Jiazi Hualian roles, appearing in more than two hundred Peking Operas. Hao absorbed the singing techniques of the Tongchui Hualian ("copper hammer painted face", painted face roles that emphasize singing), and perfectly harmonised them with his dance-acting and body postures. Hao excelled at characterization, and introduced many innovations into the Jiazi Hualian roles. It was Hao who created the "mantis eyebrow" shape as a style of painted face pattern. He also created the "monk eyes", "pink cheeks" and "bare chest" for the appearance of the character Lu Zhishen, a typical example of Jing roles, highlighting the character's forthright and down-to-earth nature.

Hao Shouchen also created new plays, including: "The Biography of Jing Ke" (Jing Ke Zhuan), "Taohua Village" (Taohua Cun), and "Niugao Gets Married" (Niugao Zhao Qin).

His representative works included: "Fighting Caobao" (Da Cao Bao), "Stealing the Imperial Horse" (Dao Yuma), "Wild Boar Forest" (Ye Zhu Lin), and "Executing Ma Su" (Zhan Ma Su).

◎ 著名演员（二）

侯喜瑞（1892～1983）：架子花脸演员。侯派特色主要体现在唱功、念白、做功、脸谱四个方面，以表现人物“精、气、神”为特征，强调“武里透文、文中有武”。

侯喜瑞8岁入科班，后拜名净黄润甫为师，得其真传。他能融合各家之长，形成自己的风格，世称“侯派”。表演矫健、细腻，注重形式美，并坚守“懂剧情、懂戏理、懂戏词”的原则。唱、念口劲严谨，字韵准确，能根据本人嗓音条件，常用沙音唱法。侯派脸谱细致、形象生动，自成一家。

侯派剧目大部分属于念、做并重。如《群英会》中的黄盖，《失街亭》中的马谡，《战濮阳》《长坂坡》《战宛城》《阳平关》中的曹操等。

代表剧目有《群英会》《长坂坡》《阳平关》《清风寨》等。

Famous Artist（Ⅱ）

Hou Xirui (1892 ~ 1983) was an actor of Jiazi Hualian ("posture painted face") roles. The features of Hou School can be discerned in the four aspects of singing, recitation, dance-acting and face-painting. Hou tried to grasp "the inner energy, manner and spirit" of characters, and emphasized that there is "civility in the martial roles, and martial arts in civil roles".

At the age of eight, Hou Xirui started his professional training. He was then taught by Huang Runfu, a celebrated actor of Jing roles. The style of the Hou School was formed by synthesizing various elements. His performance was meticulous and vigorous, and emphasizied the beauty of form. He followed the principle of understanding "the plot, the contents, and the words" of each piece. Besides adopting a rigorous approach to singing and recitation, and paying attention to rhyming patterns, there was a gritty quality to his voice. The unique face-painting techniques of Hou School became famous for their details and vividness.

Both acting and speaking skills were equally important for most of the Hou School repertoire, such as Huang Gai in "The Meeting of the Heroes" (Qun Ying Hui), Ma Su in "Loss of Jieting" (Shi Jieting), and Cao Cao in "Battle at Puyang" (Zhan Puyang), "Changban

Slope" (Changban Po), "Battle at Wancheng" (Zhan Wancheng), and "Yangping Pass" (Yangping Guan).

His representative works included: "The Meeting of the Heroes" (Qun Ying Hui), "Changban Slope" (Changban Po), "Yangping Pass" (Yangping Guan), and "Qingfeng Village" (Qingfeng Zhai).

武净

3.

武净又叫“武二花”，或“武花脸”。表演特点是重武打、扑跌，不重唱、念。要有较好的毯子功和把子功基础。武净戏最初也有些比较讲究功架、表演和念白的，如《四平山》中的李元霸、《艳阳楼》中的高登等，后来这些戏逐渐被武生行当拿去演了，并成为武生演技较为突出的作品，武净这一行反而不演这些戏了。

武净分为两类：一类重把子功架，唱、念不多，但不翻不扑，如《战宛城》中的典韦；另一类重跌打扑倒，有繁难的技巧表演，表示被对手打败，故称“摔打花脸”，如《挑滑车》中的黑风利。

武净行同其他行当一样，在京剧发展过程中也产生了不少出类拔萃的演员。特别是在19世纪末20世纪初，武戏盛行，其中的佼佼者形成了自己的风格，创立了流派，以钱金福、范宝亭最为出色。

Wu Jing (martial painted-face male role)

The Wu Jing is one of the subdivisions of the Jing roles, also known as Wu Er Hua or Wu Hualian ("the martial painted face"). Wu Jing roles put emphasis on martial skills, rather than singing and recitation. The role demands that actors have strong acrobatic and martial skills. Originally Wu Jing roles emphasized posture, acting and recitation. Good examples include Li Yuanba in "Siping Mountain" (Siping Shan) and Gao Deng in "The Tower of Yanyang" (Yanyang Lou). Over time these characters were absorbed into the Wu Sheng (martial male) roles.

Wu Jing roles can be further divided into two types: one emphasizes fighting skills, with singing and recitation but not acrobatics, such as Dian Wei in "Battle at Wancheng" (Zhan Wancheng). The other specializes in acrobatics and martial arts, and uses difficult acrobatic displays to show they are being defeated in battle. For this reason, they are also named Shuaida Hualian (defeated martial roles). Hei Fengli in

武净——《竹林记》中的余洪

"Falling Chariots" (Tiao Huache) is an example of this kind.

Similar to other role types in Peking Opera, the Wu Jing roles have produced some excellent actors. This was especially true in the late nineteenth and early twentieth centuries, when martial arts plays were popular, and the best actors formed different styles and established their own School. The most outstanding ones were established by Qian Jinfu and Fan Baoting.

◎ 著名演员（一）

钱金福（1862 ~ 1937）：京剧武净演员。自幼从于双寿、崇富贵学戏，工武净。1911 年入同庆班，与谭鑫培同台合作。钱金福武功功底深厚扎实、以勇猛见长，有大将风度。常为谭鑫培、余叔岩、杨小楼等配戏。晚年从事教学工作。

在净角脸谱的改革创新上，钱金福的贡献也是非常大的。《定军山》中的夏侯渊，从黑歪花脸改为“黑花十字”脸，以示凶恶又有帅威。他勾的脸线条匀称，无论是老脸、整脸、碎脸，都十分美丽，而且突出了剧中人的喜怒哀乐。

钱金福在扮演《定军山》中的夏侯渊、《青石山》中的周仓、《长坂坡》中的张飞、《战宛城》中的典韦等角色时，有许多创造，演出非常精彩。

代表剧目有《祥梅寺》《瓦口关》《铁笼山》《战宛城》等。

Famous Artist（Ⅰ）

Qian Jinfu (1862 ~ 1937) was an actor of Wu Jing roles. He started learning the role in his childhood, and was taught by Yu Shuangshou and Chong Fugui. Later, he entered Tongqing Troupe in 1911 and began to collaborate with Tan Xinpei. Qian excelled in martial arts skills. His performance was full of animated courage and heroic spirit, which evoked the demeanour of great generals. Qian often performed with famous actors like Tan

Xinpei, Yu Shuyan and Yang Xiaolou. In his later years, he went into teaching.

Qian's contribution to the development of the art of face-painting for Jing roles is significant. Take Xia Houyuan in "Dingjun Mountain" (Ding Jun Shan) for example. Qian changed the design of the face from a "black crooked pattern" into a "black cross-shaped pattern" to show the mightiness of the character. Qian was able to sketch well-proportioned lines to form beautiful facial make-up designs, whether it be an old face, full face, or fragmented pattern face. His innovative designs effectively highlighted the different emotions of the characters.

Qian played many Wu Jing roles, including Xia Houyuan in "Dingjun Mountain" (Dingjun Shan), Zhou Cang in "Qingshi Mountain" (Qingshi Shan), Zhang Fei in "Changban Slope" (Changban Po), and Dian Wei in "Battle at Wancheng" (Zhan Wancheng). These characters incorporated many innovations, which were exciting in performance.

His representative works included: "Xiangmei Temple" (Xiangmei Si), "Wakou Pass" (Wakou Guan), "Tielong Mountain" (Tielong Shan), and "Battle at Wancheng" (Zhan Wancheng).

◎ 著名演员（二）

范宝亭（1887 ~ 1944）：京剧武净演员。范宝亭的父亲范福泰是闻名的武花脸演员。范宝亭科班出身，自幼坐科福寿班。他武功基础扎实，长靠短打皆精。特别是跌打翻扑的特技功夫是无人能与之匹敌的。范宝亭的舞蹈化动作，既美又帅。值得一提的是，他的打法绝不是为了卖弄特技，而是按照人物和剧情的需要而有所运用。

范宝亭与众多名演员排演了不少新戏，体现了他的革新精神。如与高庆奎合演《浔阳楼》，与程砚秋合演《荒山泪》《陈丽卿》，与尚小云合演《湘江会》《娘子军》等。

代表剧目有《湘江会》《雷峰塔》《浔阳楼》《峨眉剑》等。

Famous Artist（Ⅱ）

Fan Baoting (1887 ~ 1944) was an actor of Wu Jing roles. Fan Baoting's father, Fan Futai, was also a famous Wu Jing actor. Influenced by his father, Fan Baoting started learning performance as a child, entering professional tuition (keban) with the Fushou troupe in his childhood. Fan developed a solid foundation in martial arts, and excelled in both Changkao (martial roles who wear armour) and Duanda (martial roles who fight hand to hand). In particular, he excelled at acrobatics, and no one else could match his skill. His body postures were so graceful that he almost seemed to be dancing. It is also worth mentioning that Fan's style of performance was not simply about showing off acrobatic skill; these skills were always used according to the needs of the characters and the plot.

Fan's innovations can be discerned in his rehearsal of many new plays that also featured other famous actors. For instance, Gao Qingkui co-starred in "Xunyang Building" (Xun Yang Lou); Cheng Yanqiu co-starred in "Shedding Tears on the Barren Mountain" (Huangshan Lei) and "Chen Liqing" (Chen Liqing); and Shang Xiaoyun co-starred in "The Meeting at Xiang River" (Xiangjiang Hui) and "The Army of Women" (Niangzi Jun).

His representative works included: "Meeting at Xiang River" (Xiang Jiang Hui), "Leifeng Tower" (Leifeng Ta), "Xunyang Building" (Xunyang Lou), and "The Emei Sword" (Emei Jian).

第四章　五行

丑行

丑行，京剧行当之一。丑面部化妆多在鼻梁上用白粉涂抹一个粉块，因而又叫作“小花脸”或“三花脸”。

丑面部化妆的粉块有几种形状，有方形、元宝形、倒元宝形、腰子形、枣核形等，是根据不同人物而确定的。丑扮演的人物有些是阴险狡猾、卑鄙自私的，如《望江亭》中的杨衙内、《野猪林》中的高衙内、《一捧雪》剧中的汤勤、《连升店》中的店主人等。

另外，丑扮演更多的是机警伶俐、幽默风趣，甚至正直善良的人物，即京剧传统戏里众多社会地位卑下的劳动人民，如渔夫、农民、樵夫、酒保、更夫、差役、书童、乞丐等。

从性格上看，这些人物多半是滑稽、活泼、乐观、热情的，如《女起解》中的崇公道，是一位忠厚、善良，有正义感的老差役，《秋江》中的艄公，是一位助人为乐、风趣的老人。

最突出的是《徐九经升官记》剧中的徐九经。他的表演是滑稽可笑的，他的容貌表面上有点丑，但是他宁可牺牲个人利益，坚决顶住权势的威逼，勇敢为百姓伸张正义，心灵确实是真、善、美的。

从表演技艺来看，丑可分为文、武两类。

Chou

Chou (clown) is one of major role types in Peking Opera. It can be easily identified by a white patch across the bridge of the nose. Consequently, people often call it Xiao Hualian (little painted face) or San Hualian (third painted face).

The make-up for the Chou appears in many different shapes: square-shaped, shoe-shaped, gold ingot-shaped, kidney-shaped, date seed-shaped – each referring to a different kind of character. Some Chou roles represent treacherous, cunning, selfish and contemptible characters, such as Yang the Official's Son in "The Riverside Pavilion" (Wangjiang Ting), Gao the Official's Son in "Wild Boar Forest" (Ye Zhu Lin), Tang Qin in "A Handful of Snow" (Yi Peng Xue) and the Innkeeper in "Liansheng Inn" (Liansheng Dian).

However, the Chou is more likely to portray humorous, morally upstanding, sharp-witted and kind characters, most of whom are from the lower working classes (such as fishermen, peasants, woodcutters, bartenders, night watchmen, servants, attendants, beggars and so on).

Most characters are funny, lively, optimistic and warm. Chong Gongdao, an old Yamen runner in "A Woman Escorted for Trial" (Nü Qijie) is a courteous and kind-hearted person. The Boatman in "Autumn

River" (Qiu Jiang), is a humorous role who finds pleasure in helping others.

The most prominent Chou character is Xu Jiujing in "Xu Jiujing's Promotion" (Xu Jiujing Sheng Guan Ji), whose performance is hilarious. Although he seems to be a buffoon on the surface, he sacrifices his own personal interests to speak out against the abuse of power and stand up for the ordinary people. His heart pursues truth, goodness and beauty.

In performance, the Chou role can be further divided into Wen Chou and Wu Chou.

文丑

1. 文丑表现的人物并非全是文人，只是表演上不重武功，是相对于武丑而言的。文丑又划分为方巾丑、袍带丑、茶衣丑、褶子丑等。

Wen Chou (civilian clown)

Wen Chou (civilian clown) is a subcategory of the Chou role. The characters who are portrayed by the Wen Chou are not necessarily all educated members of the literati; rather these roles do not require skills in martial arts. Wen Chou can be further divided into the Fangjin Chou ("kerchief" chou), Paodai Chou ("belt robe" chou), Chayi Chou ("tea garment" chou) and Xuezi Chou ("casual coat" chou).

方巾丑

1.1.

方巾丑多扮演一些达官贵人，有文化的人和有一定身份的上层社会人士。方巾丑头戴方巾，身穿褶子，有的手里还拿把扇子，迈着四方步，举止缓慢文雅，文质彬彬，但带点迂腐的酸气，这类人擅长察言观色、阿谀奉承、吹牛拍马、诡计多端。这类角色在戏曲舞台上一般是反面人物，通过愚笨或狠毒的形象，反衬正面人物形象的睿智与正义，一般说韵白，如《蒋干盗书》中的蒋干、《乌龙院》中的张文远等。

代表作品有《蒋干盗书》《乌龙院》等。

Fangjin Chou ("kerchief" chou)

Fangjin Chou ("kerchief" chou), refers to officials, members of the literati and noble lords. On stage, the character normally wears a square-shaped kercheif and carries a fan in his hand, walking in measured steps with gentle gestures and slow movements, but an unpleasant air. These movements convey that these characters are conceited petty scholars who have a pedantic disposition. They are skilled in observing, flattering, bragging, and scheming. These roles are usually negative characters, stupid and wicked, and serve as a contrast to the wise and righteous hero. These characters generally speak in yunbai (a form of stylized speech which gives the character an air of decorum). Examples of this role include Jiang Gan in "Jiang Gan Steals a Letter" (Jiang Gan Dao Shu), and Zhang Wenyuan in "Black Dragon Courtyard" (Wu Long Yuan).

Representative works："Jiang Gan Steals a Letter" (Jiang Gan Dao Shu) and "Black Dragon Courtyard" (Wu Long Yuan).

方巾丑——《蒋干盗书》中的蒋干

◎ 著名演员

萧长华（1878～1967）：文丑演员。艺名宝铭，号和庄。江西人，生于北京。为著名京剧丑行表演艺术家，创立了丑行流派，师承宋万泰，又吸取黄三雄、刘赶三等名角的艺术精华并有所创新，形成丑角影响力最大的流派，世称“萧派”。他在舞台上演出近七十年。尽管他是一名丑角演员，但他的学生在生、旦、净其他行当里也很活跃。他擅长用严肃、深刻的手法摹尽世态人情，揣摩人物，表现人物的冷峻幽默，尤擅方巾丑。

Famous Artist

Xiao Changhua (1878 ~ 1967) was a famous Wen Chou actor, whose artistic name was Baoming, and whose alternative name was Hezhuang. He was born in Beijing but was a Jiangxi man. A famous Chou actor, he is regarded as the founding father of the Chou Schools. Xiao studied under the actor Song Wantai, and absorbed further techniques from Huang Sanxiong and Liu Gansan. Through his own innovations, he created the most influential acting School for Chou roles: the Xiao School. His career lasted for almost seventy years. Although he was a Chou actor, he also taught students from the Sheng, Dan and Jing roles. He applied his craftsmanship deeply and seriously to demonstrate the complexity of the human condition, to interpret and portray the characters in every conceivable way, showing how characters that lack spirit are incomplete, cold, and humorous. He was especially good at the Fangjin Chou.

方巾丑——《活捉三郎》中的张文远

袍带丑

1.2.

袍带丑又称官衣丑，身穿官服，腰围玉带，头戴圆翅纱帽，正反面人物均有。袍带丑有时饰演帝王将相、公卿大夫等喜剧人物；有时饰演正直又机智的角色，如《蝴蝶杯》中的董温；有时饰演怯懦而昏庸的角色，如《湘江会》中的齐景公；有时饰演贪官污吏或者卑鄙无耻的势利小人，如《审头刺汤》中的汤勤；还饰演自作聪明的人物，如《春草闯堂》中的胡进、《四进士》中的刘题、《六月雪》中的山阳知县等。

代表作品有《春草闯堂》《审头刺汤》《蝴蝶杯》《湘江会》《四进士》《六月雪》。

Paodai Chou (“belt robe” chou)

Paodai Chou ("belt robe" chou), also known as Guanyi Chou ("official robes" chou), wears a jade belt around his waist, and a round-winged gauze cap on his head. The Paodai Chou refers to both positive and negative characters. The role sometimes portrays humorous emperors, senior officials and doctors. Sometimes they play straight, intelligent characters, such as Dong Wen in "The Butterfly Cup" (Hudie Bei). Sometimes they play cowardly and stupid characters, such as Emperor Qi Jinggong in "A Meeting at Xiang River" (Xiangjiang Hui). Sometimes they play corrupt or shameless villains, such as Tang Qin in the play "Assassinating Tang for the Approval of a Head" (Shen Tou Ci Tang). Other examples include Hu Jin in "Chun Cao Bursts into the Hall" (Chuncao Chuang Tang), Liu Ti in "The Lawsuit" (Si Jinshi), and the Magistrate of Shanyang County in "Snow in Midsummer" (Liu Yue Xue).

Representative works: "Chuncao Bursts into the Hall" (Chuncao Chuang Tang), "Assassinating Tang for the Approval of a Head" (Shen Tou Ci Tang), "The Butterfly Cup" (Hudie Bei), "A Meeting at

袍带丑——《昭君出塞》中的王龙

Xiang River" (Xiangjiang Hui), "The Lawsuit" (Si Jinshi) and "Snow in Midsummer" (Liu Yue Xue).

1.3.

茶衣丑

茶衣是一种蓝布短褂子，是戏曲服装中最简陋的戏衣，属于古代劳动者专用的服装，所以茶衣丑扮演的都是劳动人民。角色形象数量众多，多以配演为主。这些人物个性鲜明，造型丰富，有忠厚老实、热情善良的好人，也有不务正业、好逸恶劳的恶人。茶衣包含两类：一类是茶衣，短的蓝布褂子，一般是对襟；另一类是大袖，一般系在旁边。茶衣和大袖系一个蓝色的或白色的小腰包，术语叫茶衣腰包。如《秋江》中的艄公、《芦中人》中的渔翁、《问樵闹府》中的樵夫、《骆马湖》中的酒保、《打樱桃》中的书童等。

代表作品有《秋江》《问樵闹府》《芦中人》《打樱桃》。

Chayi Chou (tea-garmented chou)

Chayi Chou (tea-garmented chou) is one kind of Wen Chou. The Chayi is a kind of hip length garment made of blue cloth with black sleeves that used to be worn by the working classes. Thus the characters played by the Chayi Chou are also all from the working classes. There are many different kinds of characters, and they appear in many performances. Each character is distinctive, some are honest and kind, whilst others are idle and wicked. There are two types of Chayi, one is a short blue casual coat with an opening at the centre of the front; the other has long sleeves, which usually are tied up to the side. A blue or white short apron-like skirt called Yao Bao is worn outside the trousers.

茶衣丑——《秋江》中的艄翁

Examples of Chayi Chou can be found in roles such as The Boatman in "Autumn River" (Qiu Jiang), The Fisherman in "The People of the Reeds" (Lu Zhong Ren), The Woodcutter in "Making a Disturbance in Ge's Mansion" (Wen Qiao Nao Fu), and The Bartender in "Luo Ma Lake" (Luo Ma Hu).

Representative works： "Autumn River" (Qiu Jiang), "Making a Disturbance in Ge's Mansion" (Wen Qiao Nao Fu), "The People of the Reeds" (Lu Zhong Ren) and "Picking Cherries with Sticks" (Da Yingtao).

1.4.

褶子丑

褶子丑说京白，有时穿长衫或箭衣，表演风格比茶衣丑略严谨，如《连升店》中的店主人、《苏三起解》中的崇公道等。

代表作品有《苏三起解》《连升店》等。

Xuezi Chou ("casual coat" chou)

Xuezi Chou ("casual coat" chou), speaks in jingbai (a style of speech similar to the Beijing dialect). The Xuezi Chou usually wears a long gown and an archer's uniform, whose performance style is more rigorous than the Chayi chou. Examples include The Inn Keeper in "Liansheng Inn" (Lian Sheng Dian), and Chong Gongdao in "Susan Setting Out for Trail" (Susan Qi Jie).

Representative works: "Liansheng Inn" (Lian Sheng Dian), and "Su San Setting Out for Trail" (Susan Qi Jie).

褶子丑——《苏三起解》中的崇公道

老丑

1.5.

介于方巾丑与茶衣丑之间的角色。穿长衫或者箭衣，带髯口，头戴白毡帽。表演风格比茶衣丑更为严谨，以念白为主，说京白，有时也念方言，多扮演心地善良、性格诙谐的老年男子。在化妆上与其他丑角类别不同，在脸上勾画豆腐块的同时，在额头、眼角处勾画皱纹，还勾画白眉毛，突出老年人的年龄感。老丑所带的髯口多为“白五嘴”。

老丑塑造的角色，为社会下层的平民百姓。如《乌盆记》中以卖草鞋为生的张别古、《问樵闹府》中的砍柴樵夫。这类角色多穿老斗衣或大袖腰包，头戴白发鬏儿，以表现平民百姓的朴素本色。

代表作品有《乌盆记》《问樵闹府》。

Lao Chou (elderly chou)

Lao Chou (elderly chou) is in-between the Fangjin Chou and Chayi Chou. The Lao Chou wears a long gown or archer's uniform with rankou (artifical beard) and white felt cap. The performance style of the Lao Chou is more rigorous than the Chayi Chou. Recitation is very important, usually undertaken in jingbai (a style of speech similar to the Beijing accent), though sometimes the character speaks in other dialects. The Lao Chou usually portrays kind and humourous elderly men. Unlike other kinds of Chou role, the Lao Chou has a painted patch of white across his face that looks somewhat like a block of tofu, with the addition of wrinkles on the brow and in the corners of the eyes. The whitening of the eyebrows also suggests that this is an elderly role. The Lao Chou usually wears a beard called a "bai wu zui" (literally "white five strands").

The Lao Chou role usually depicts people of the lower class. They usually wear clothes that are conventionally associated with old men called lao dou yi (or gowns with long sleeves). Examples include Zhang

老丑——《问樵闹府》中的樵夫

Bie Gu in "The Injustice" (Wupen Ji), and The Woodcutter in "Making a Disturbance in Ge's Masion" (Wen Qiao Nao Fu).

Representative works： "The Injustice" (Wupen Ji), and "Making a Disturbance in Ge's Mansion" (Wen Qiao Nao Fu).

2.

武丑

武丑又名“开口跳”，扮演身怀绝技、武功不凡的江湖武林侠客。在扮演有些人物时，除了擅长蹿蹦跳跃的功夫以外，还要善于念白，突显口齿伶俐，达到清脆悦耳的念诵效果。

如《三岔口》中的刘利华、《挡马》中的焦光普、《九龙杯》中的杨香武、《连环套》中的朱光祖、《雁翎甲》中的时迁等。

代表作品有《三岔口》《挡马》《九龙杯》《连环套》等。

Wu Chou (martial chou)

Wu Chou (martial chou), also known as kao kou tiao ("open the mouth and jump"), performs plays that require special martial arts skills. In addition to being good at acrobatics and martial arts, actors must also be good at recitation. These skills in recitation serve to show through dialogue how the character is quick-witted. For this reason, the actor must recite lines with a crisp and pleasant voice. Examples of Wu Chou roles can be found in: Liu Lihua in "At the Crossroads" (San Cha Kou), Jiao Guangpu in "Stopping the Horse" (Dang Ma), Yang Xiangwu in "The Nine Dragon Cup" (Jiu Long Bei), Zhu Guangzu in "The Stratagem of Interlocking Rings" (Lian Huan Tao), and Shi Qian in "The Goose-Feather Armour" (Yan Ling Jia).

武丑——《三岔口》中的刘利华

Representative works: "At the Crossroads" (Sancha Kou), "Stopping the Horse" (Dang Ma), "The Nine Dragon Cup" (Jiulong Bei) and "The Strategem of Interlocking Rings" (Lian Huan Tao).

◎ 著名演员

叶盛章（1912 ~ 1966）：武丑演员。原名焕，字耀如。安徽太湖人，出生于北京。生于梨园世家，武戏学王长林，文戏则学萧长华、郭春山。他的念白爽利，武打干净。在艺术上勇于革新，在传统剧目的改革过程中，创造了许多新特技，形成了自己的艺术风格。

Famous Artist

Ye Shengzhang (1912 ~ 1966) was a famous Wu Chou actor. His original name was Huan, and his alternative name was Yaoru. He was born in Beijing, but his home was in Taihu county, Anhui province. He came from a family of performers, and learnt Wu Chou (martial clown) plays from Wang Changlin and Wen Chou (civil clown) plays from Xiao Changhua and Guo Chunshan. His recitation was clear and his martial artistry was neat. By reforming the traditional repertoire, he created many new stunts. These innovations established his own artistic style.